DUE
P9-AFO-890

Harvard Economic Studies

Volume 161

Awarded the David A. Wells Prize for the year 1986 and published from the income of the David A. Wells Fund.

Industrial Restructuring with Job Security

The Case of European Steel

Susan N. Houseman

Harvard University Press
Cambridge, Massachusetts
London, England
1991

Printed in the United States of America
10 9 8 7 6 5 4 3 2 1

This book is printed on acid-free paper, and its binding materials have been chosen for strength and durability.

Library of Congress Cataloging-in-Publication Data

Houseman, Susan N., 1956–
Industrial restructuring with job security: The case of European steel / Susan N. Houseman.
p. cm.—(Harvard economic studies; v. 161)
Includes bibliographical references and index.
ISBN 0–674–45175–9
1. Steel industry and trade—European Economic Community countries. 2. Iron and steel workers—European Economic Community countries—Supply and demand. 3. Job security—European Economic Community countries—Case studies. I. Title. II. Series.
HD9525.E8H68 1991 90–44471
331.12′9691′094—dc20 CIP

In memory of my father,
Harry A. Houseman

Preface

I began studying the steel industry in the context of trade disputes between the United States and European Community (E.C.) countries over the alleged dumping of European steel onto U.S. markets. The industries on both continents were suffering from low capacity utilization and other structural problems. American producers were accusing European producers of trying to expand their market shares abroad to compensate for weak markets at home.

Although my initial work focused on trade issues and strategies for restructuring the industry, I became increasingly interested in labor adjustment problems. Certainly, underlying the trade conflicts over steel between the United States and Europe (and among the European Community countries themselves) was the issue of jobs. A major reason that restructuring in steel industries had become such a controversial policy issue on both continents was the potential dislocation to workers and their communities. I felt that in order to understand these trade disputes and the industry strategies for restructuring, I should first address the problems associated with work force reductions.

The steel industry epitomizes tensions in many industrialized countries over, on the one hand, the need to restructure their industrial bases to reflect shifting trade and demand patterns, and, on the other, pressures to provide stable employment and to avert massive dislocation of workers and their communities. In the United States, large-scale layoffs in industries like steel resulted in increased worker demands for job security provisions in collective agreements and legal protections against layoffs. In 1988 the U.S. Congress passed for the first time legal requirements that employers provide advance notice to workers in the event of a mass layoff or plant closure. In many European countries, where private sector practices of strong job security and laws regulating layoff practices were well established, the 1980s brought pressure from business to relax job security in order to facilitate rapid restructuring.

In this book, I look at efforts in certain countries to combine strong job security with rapid restructuring, through a case study of the E.C. steel industry. When I went to Brussels to study restructuring under the Community's steel plan, I realized that the E.C. steel industry offered a natural experiment for examining the effects of job security on how an industry adjusts to structural change. The steel industries in member countries faced similar problems in the 1970s and 1980s: a decline in demand for steel, excess capacity, and, consequently, poor profit performance. Moreover, restructuring strategies to modernize and reduce capacity were coordinated by the European Community, and I was able to use detailed, comparably defined data at the plant level as a basis for the analysis in the book. However, the methods of work force reduction and the level of job security afforded workers during the restructuring process varied considerably across countries. In many continental countries, restructuring was accomplished with minimal recourse to layoffs, whereas in Britain, restructuring was accompanied by massive layoffs. In this book I examine the factors behind the strong job security in certain countries and the effects that it had on employment, production, and investment decisions in the steel industries of those countries.

In the book I do not make arguments to advocate, on efficiency grounds, either stronger or weaker job security for workers. Rather, the book's analysis and conclusions emphasize distributional issues. The job security that workers have reflects the underlying legal and implied rights they have in their jobs; it also reflects how the costs of adjustment to economic change are distributed in society. A recurrent theme of this book is that the way in which rights in society are allocated—or how these adjustment costs are distributed—affects the way in which industries optimally respond to changing economic conditions, both in the short run, through production and employment decisions, and in the long run, through investment decisions that affect the configuration of the industry. In the book, I point out some of these effects and the economic tradeoffs that strong job security for workers implies.

The book incorporates two quite different empirical research styles. The analysis is based largely on statistical analysis of plant-level data. I also conducted extensive interviews with government, labor, and company representatives in the seven countries. From these interviews I collected information on work force reduction programs and their costs in the steel industries in these countries—findings which I report in the book. I also gained useful insights into the perceptions of the various parties to the restructuring programs as to why certain choices were made and what the major obstacles to work force reductions were. Although the contributions that these interviews made to this book are in many respects intangible, they were important in shaping and providing confirmation for many of the ideas presented in it.

I collected data on the steel industry of seven E.C. countries during two trips to Europe in 1984 and 1986. I am grateful to many individuals at the European Commission who provided me with information about the European Community's steel policies and with contacts in the steel industry. Among these individuals were David Brew, Pierre Defraigne, Aldi Farber, Jacques Soenens, Pierro Squartini, Guido Vanderseypen, and Herbert Von Bose. During my trip in 1986, the U.S. embassies in Belgium, France, West Germany, and the Netherlands were quite helpful in lining up appointments and providing translation services, where necessary. For assistance with my fieldwork, I am indebted to many company, labor, and government representatives, too numerous to list here, who freely gave me their time.

I received financial support for travel in Europe from the Krupp Foundation at the Center for European Studies, Harvard University, and from the University of Maryland. While I was in Brussels, on both occasions, Herman Daems generously supplied me with office space at the European Institute for Advanced Studies in Management.

I am deeply indebted to Richard Caves for his valuable comments at virtually every stage of the research and for his encouragement and support as this book took shape. I also benefited from discussions with many other individuals while researching and writing this book. In particular, I wish to thank Katharine Abraham, Roger Blanpain, David Bloom, Aldi Faber, Richard Freeman, Dan Hamermesh, Doug MacLean, Leo Sleuwaegen, and Guido Vanderseypen. I would also like to thank Raymond Ascoberta and Robert Verulund for their excellent research assistance; Carolyn Thies and Claire Vogelsong for their assistance with proofreading the manuscript; and Maria Ascher for her help in editing the manuscript.

Finally, my husband, Curtis, deserves special thanks. I worked on writing this book over the course of two years, during which time our daughter, Evelyn, and son, Michael, were born. Certainly, all of the support and good humor that Curtis provided at home were invaluable during these very special and busy years.

Contents

Industrial Restructuring with Job Security

1

Introduction

Job security for workers has various effects on the way in which firms in an industry adjust to a decline in demand and to a situation of excess capacity. This book presents a study of these effects, through an analysis of the steel industry in European Community (E.C.) countries. Production of crude steel by E.C. countries in 1986 was almost 30 percent lower than the peak output in 1974. During the same period, employment in the E.C. steel industry fell by almost 50 percent. The reasons for the severe decline include a drop in the demand for steel, and rapid product and process innovations that left the European steel industry uncompetitive in world markets. The problems faced by the E.C. steel industry, though extreme, are mirrored in numerous sectors in industrialized countries.

One consequence of the decline of important industries such as steel has been a major shift in the composition of jobs in these countries and widespread worker dislocation. With the growth of international trade, the development of Third World countries, and rapid technological progress, such structural change and its effect on jobs in industrial economies is unlikely to diminish in the near future.

Plant closures and massive layoffs result in personal, social, and economic costs to management, workers, and their communities. In the United States, widespread dislocation from the decline of major manufacturing industries in the 1970s and 1980s underlies the recent surge in job security provisions in collective agreements, political pressures for trade protection, and the enactment of legislation requiring advance notice prior to mass layoffs. These developments have helped establish greater worker rights to job security and have helped shift some of the costs and risks of structural change onto companies or the public sector.

In Europe, public and private sector policies supporting job security were well established by the 1970s. Implicit and explicit worker rights, such as the right to receive advance notice prior to dismissal or layoff, have a long history in Europe. Restrictions on dismissals and layoffs were greatly strengthened in employment security laws in the 1960s and 1970s in many European countries. In addition to legal restrictions, labor practices adopted by the private sector have tended to shield workers from changes in production levels to a far greater degree than is common in the United States.

Job security is appealing to many people on equity grounds. Proponents argue that certain groups of workers should not bear the brunt of economic change. Proponents also argue that greater worker rights to job security will reduce worker resistance to change and thereby facilitate adjustment to the natural and continual changes occurring in any economy. Opponents have offered contrary arguments: job security will impede restructuring in declining sectors and inevitably will weaken the ability of these industries to compete in world markets.

Numerous studies in recent years, particularly in the United States, have documented the costs to workers of being permanently laid off as a result of a structural decline in demand.[1] In this book, I turn that question around. If workers have some protection against layoff, how do their rights to job security affect the way an industry adjusts to a permanent decline in demand for its product?

Restructuring in the E.C. steel industry offers a particularly good case for studying this issue. Job security has been a central feature of restructuring programs developed at the Community, state, and company levels. Steelworkers in all countries enjoyed substantial protection against layoff through Community laws, job security legislation in individual member states, and, most important, contractual rights in their collective agreements. Steel companies in many countries achieved large cuts in capacity and work force levels, for the most part without layoffs. However, there were significant differences across countries in the degree of job protection afforded workers. Exploiting differences in job security across countries, I examine how job security affected employment, production, and investment decisions.

My empirical analysis uses detailed data at the plant level, obtained from the Commission of the European Communities (the administrative arm of the European Community, which played an active role in coordinating member state policies and regulating production, prices, and investment in steel). The data cover eighty-eight steel works that represent the principal integrated producers in the seven major E.C. steel-producing countries: West Germany, France, Britain, Italy,

Belgium, Luxembourg, and the Netherlands. These companies account for about 80 percent of steel capacity in the European Community.

In addition, I collected detailed information on the methods of work force reduction in each country. Most of this information was obtained from interviews, conducted in 1984 and 1986, with employer, worker, and government representatives in each of the seven countries. From the interviews, I obtained information on the principal methods of work force reduction used, any changes in work force reduction programs over the period, and the level of job guarantee afforded workers. I was particularly interested in whether companies had used layoffs, and, if so, the type and level of benefits received by affected workers.

The interviews also were a means for understanding how individuals involved in the restructuring process perceived the problem of labor adjustment and the issue of job security. Whenever possible, I spoke with employer, employee, and government representatives in each country.[2] Interviewing representatives of all parties concerned with the negotiations over work force reductions served to confirm the "facts" of how these reductions were accomplished. It also provided insights into the bargaining process and into people's perceptions of why certain alternatives were selected and what the most important problems in reducing the work force were.

The book covers restructuring in the steel industry from just prior to the onset of the crisis in 1974 through 1986. Setting the stage, Chapter 2 provides background information on the causes of excess capacity in the E.C. steel industry, on E.C. policy for the steel industry, and on restructuring strategies in the individual member states. The E.C. steel industry has undergone restructuring involving substantial cuts in production, capacity, and employment in a relatively short period of time. That restructuring has involved an exceptional degree of coordination between steel companies and member states at the Community level. Following an initial period of price and subsidy wars, a series of E.C. agreements, collectively known as the Davignon Plan, governed adjustment in steel. Short-term measures included quotas on key product groups over the adjustment period in order to stabilize prices and company profits. Longer-term measures involved the negotiation of capacity reductions on a company-by-company basis to bring capacity in line with production by 1986. Although excess capacity still exists in certain segments of the industry, the Davignon Plan largely achieved its goal.

The remainder of the book focuses on labor considerations and the ways in which worker rights to job security affect the restructuring process. Chapter 3 describes the methods of work force reduction adopted in each country in light of worker rights in jobs and the dependence of many regions on steel for employment. The work force reduction programs in the E.C. steel industry were a product of historical, economic, and political factors. Traditionally, steel companies had provided workers with stable employment. The economic dependence of many steel regions on that industry and the geographic immobility of steelworkers lowered the opportunity costs of labor in steel and raised political opposition to layoffs. The decline in the steel industry also coincided with a movement in many Community countries to increase legal protection against layoffs. The confluence of these factors resulted in quite strong guarantees of employment in collective agreements, often backed by financial assistance from governments. The notable exception was Britain, where an integral part of the Thatcher government's domestic policy was to break organized labor's power in the economy; its nationalized steel industry became a test case.

The extent of the job security afforded workers was clearly reflected in the methods of work force reduction used. On the Continent, steel companies relied extensively on attrition, early retirement, job buyout programs, and various measures to cut working time in order to reduce labor input. In contrast, the British steel industry achieved work force reductions through massive layoffs. However, even in Continental countries tensions over the high costs to companies and governments of using alternatives to layoffs resulted in some compromises on job security.

My fieldwork for this study, reported in Chapter 3, heavily influenced my theoretical treatment of work force adjustment, developed in Chapter 4. In the interviews I conducted, company, worker, and government representatives—without exception—stressed the strong attachments workers had to job and community and the resulting problem of labor immobility. Often, local economies were heavily dependent on steel for employment. Workers affected by plant closures or mass layoffs had to move, or would risk long-term unemployment. As evidence of worker opposition to moves, collective agreements typically protected workers against transfers out of their region, in countries where job security was strong. Studies of worker dislocation in Europe as well as in the United States likewise have stressed problems of immobility.[3] Certainly workers do have attachments to job and community that make them less mobile and that influence economic choices. The assumption underlying my discussion here is that these attachments are important.

Chapter 4 draws on neoclassical economic theory and behavioral theory to provide a foundation for understanding the way in which worker rights to job security affect adjustment in a declining industry. The notion that workers have strong attachments to job and community that curtail their mobility is incorporated into the formal models and theoretical discussion. If workers are interested only in maximizing their income, then rights assignments in jobs do not affect adjustment in a declining industry. This assumption about worker utility is implicit in the Coase Theorem, which states that resource allocation is independent of the assignment of property rights in society. If, however, workers value certain locational or other attributes of their jobs independent of the income their jobs provide, then the assignment of rights in jobs will influence resource allocation. Although the explanation for this phenomenon differs between neoclassical and behavioral theories, all of the theories predict that the effective price of labor in a declining industry will vary according to how rights in jobs are assigned, or, alternatively, how the costs of adjustment are divided, between workers and firms. The price of labor, in turn, affects optimal employment, production, and investment decisions in the industry.

There is a tendency in the economics literature to try to separate "distributional" issues from "efficiency" issues. If a unique, efficient equilibrium exists, then the policy prescription is straightforward. One selects the policy resulting in the efficient allocation of resources and accomplishes the desired income distribution through transfers. The point developed in Chapter 4 and emphasized subsequently in the book is that such a unique equilibrium may not exist, because worker rights in jobs affect how firms in the industry respond to structural change; the distribution of income cannot be neatly separated from the allocation of resources.

Tests of the effects of job security on restructuring in the E.C. steel industry are contained in Chapters 5 and 6. Empirically, there is a strong link between steelworkers' rights to job security in a particular country and the adjustment of employment levels in plants in that country. Controlling for the possibility that causality runs in the other direction—that job security might have been relatively strong in certain countries because there was less need to restructure—empirical evidence presented in Chapter 5 suggests that stronger job security resulted in substantially smaller decreases in employment levels. However, companies in countries with strong job security tended to adjust average hours worked by their employees more through the reduction of overtime hours, the use of short-time work, increased vacation, or the permanent reduction of the work week. In countries with strong job security, greater adjustment of average hours worked compensated in part for the lack of adjustment of work force levels in the short run.

Although job security most directly affects the adjustment of employment and hours at the plant level, it also has more pervasive impacts on restructuring strategies. The larger the proportionate reduction in output, the more costly it is to avoid layoffs and handle labor reductions entirely through attrition and the reduction of average hours. By increasing the costs of reducing work force levels, stronger job security would be expected to result in greater equalization of the percentage reduction in production and capacity across plants.

Chapter 6 reports my findings on these issues. One consequence of the Davignon Plan was that it helped preserve initial production and capacity shares, a solution that arguably was due in part to the strong job security in many Community countries. Moreover, when one compares across countries, one finds that the level of job security had a substantial impact on companies' ability to close plants and concentrate production at sites with more efficient equipment and locational advantages.

Therefore, the level of job security reflected in work force reduction programs in the member states went beyond affecting short-run productivity and the choice between the reduction of employment and the reduction of average hours. By influencing production, investment, and closure decisions, it has had more lasting effects on the structure of the steel industry in Europe.

There are important qualifications to the general conclusion that job security hampers restructuring. The steel industries in certain countries, notably in Luxembourg and France, achieved relatively large work force reductions and productivity gains despite strong job security. In addition to job buyout schemes and measures to reduce working time, companies in Luxembourg, France, and elsewhere instituted effective systems of internal and external transfers, the subcontracting out of excess labor, and greater flexibility on the scheduling of hours. These measures typically required that labor make some concessions concerning work rules in return for job guarantees, and facilitated rapid restructuring with strong job security.

The empirical evidence presented in the book may be read by some as an argument against job security for workers. It is not. Although mistakes no doubt were made during restructuring of the E.C. steel industry, this book does not test the efficiency of specific policies.

Rather, the main argument of the book is that worker rights in jobs have important implications for how the costs of adjustment are distributed in society and for how industries respond to economic change. The effects that strong worker rights to job security have on restructuring in a declining industry are not, in themselves, inefficient in the Pareto sense that someone could be made better off without making anyone else worse off. From a social perspective, worker rights to job security have costs, which include lower productivity growth. They also have social benefits, which include community stability and a more equitable distribution of the risks and costs of economic change in society. In this book, I try to elucidate these effects.

2

The Response to Excess Steel Capacity in the European Community

The severe problem of excess capacity in the E.C. steel industry was unanticipated. Steel production in Europe reached record levels in 1974, and most industry analysts expected the high growth of the postwar years to continue. Consequently, the downturn in 1975 precipitated by the first wave of oil price increases was perceived as being cyclical in nature; companies continued ambitious expansion plans, positioning themselves for the expected upturn. The structural component of the crisis was not widely recognized for several years.

As low prices and surplus capacity persisted, many firms faced staggering financial losses. Rather than allowing major segments of their industries to go bankrupt, governments—particularly in France, Britain, Belgium, and Italy—began to subsidize their steel industries heavily. The subsidies, which were illegal under E.C. law, only exacerbated the crisis by enabling companies to operate rather than close their marginal plants. Their net result was to bring prices down further, and even the more competitive firms began to experience serious financial difficulties. Other member states retaliated with their own subsidies and with threats to close their borders to intra-E.C. steel imports.

In response to the rapidly deteriorating common market in steel, the Community adopted a set of policies to govern production, prices, investment, and capacity reduction. These policies became known, collectively, as the Davignon Plan. The Davignon Plan involved an exceptional degree of coordination among steel producers, member state governments, and the European Commission. Its goal was to end subsidization to and excess capacity in the industry by 1986.

By 1986 the target for capacity reduction had been met, and controls over prices and production were being phased out. In 1988 these controls were eliminated, though the appreciation of the European currencies against the dollar in the late 1980s further weakened the competitiveness of the industry and posed a new threat of surplus capacity.

This book covers restructuring under the Davignon Plan. Although considerable restructuring remains to be done in certain countries, the Davignon Plan has had a profound impact on how restructuring has been accomplished in the E.C. steel industry. Arguably, it has set a precedent for future industrial policy in steel and other sectors.

The European Coal and Steel Community

The industrial policies developed by the European Community and member states to deal with the crisis were heavily influenced by existing institutions and historical precedent. In devising and implementing restructuring plans, the oligopolistic market structure of the steel industry and the special supranational authority of the European Community over the steel sector facilitated an exceptional degree of coordination between producers, member state governments, and the European Commission.

The European Coal and Steel Community (ECSC), the predecessor of the European Community, originally had six members: France, West Germany, Belgium, Luxembourg, the Netherlands, and Italy. Although by 1974 the European Community had expanded to include ten Western European countries, of the new members only Britain had a steel industry of significant size.[1] The Treaty of Paris, signed in 1951, established the ECSC. The Treaty of Rome, signed in 1957, extended the common market to all sectors. While ECSC institutions have been integrated with those of the European Community, the Treaty of Paris still governs the coal and steel sectors. This legal arrangement is important, for key supranational powers were diluted in the Treaty of Rome.

The supranational authority vested in the ECSC reflects the political climate in Europe in the aftermath of World War II. The Ruhr, Lorraine, and Saar regions, with their concentration of coal and steel industries, formed the "industrial heartland" of Western Europe. Dominated by Germany, the area had constituted the basis for German economic and military power. France, which had been invaded by Germany three times in less than a hundred years, feared German remilitarization, and thus opposed establishment of an independent German state after World War II. The formation of the ECSC was, in part, a *quid pro quo* for French recognition of Germany. Many people believed that integrating the coal and steel sectors, with supervision and control by an international authority, would prevent remilitarization and thus the outbreak of war in Europe in the future.

ECSC institutional arrangements reflect not only political factors, but also the market organization in the coal and steel sectors. Powerful cartels had a long history in both industries. The stated rationale for various institutional arrangements reveals a curious tension between two opposing views: one, that the cartels were too strong and must be controlled to protect the consumer; the other, that excessive competition existed and cooperation should be promoted. Three areas of E.C. jurisdiction in steel that are of particular relevance to the recent restructuring in that industry are pricing; investment; and special emergency powers to control prices, production, and distribution in the event of a crisis.

ECSC pricing rules establish a system of open delivered prices and prohibit price discrimination. Though the pricing rules ostensibly were designed to protect the consumer, many people have noted the similarity of the arrangement to multiple basing-point systems set up by private oligopolies. The system renders the pricing structure extremely transparent, and in this way facilitates oligopolistic collusion. Furthermore, the prohibition of price discrimination helps firms protect traditional markets. A producer cannot undercut prices of other steelmakers in order to attract customers in new markets without cutting prices for customers in its traditional markets. The cost of competing for new markets, therefore, is high.

Interestingly, when the ECSC was formed, analysts were predicting excess capacity in the European steel industry. Steel industries were being rapidly rebuilt following the war, and it was feared that excess capacity and cutthroat competition would result.[2] Instead, the market displayed unexpectedly strong growth, and excess capacity did not develop.

Reflecting that initial, inaccurate prediction, the Treaty of Paris includes measures giving the Community some powers in the area of investment. Under the treaty, states are prohibited from subsidizing their coal and steel industries. The European Community has limited power to control investment. It can provide some funds itself for investment, as well as for technical and economic research and readjustment assistance. Although it cannot block firms from making investments, it can restrict their access to outside funding. Historically, its most important function has been to collect and disseminate information on the steel industry. The European Community requires all firms to submit investment plans. Using this information, it analyzes market and price trends, develops forecasts, and draws up general objectives concerning modernization and capacity.

Although the development of long-term investment strategies is designed to prevent bottlenecks and surplus capacity, the E.C. has the authority to invoke powerful short-term measures to deal with an immediate crisis. In the case of serious shortage, it can establish a distribution system and maximum prices. In the event of surplus capacity, the E.C. can establish minimum prices and, with the approval of the Council of Ministers (that is, with the approval of all of the national governments), it can declare a state of "manifest crisis" and introduce production quotas. The E.C. first exercised its powers under the manifest crisis regime in 1980.

With respect to external relations, member states, while under the Treaty of Rome, must abide by a common trade policy; but under the Treaty of Paris, they have the right to set their own tariffs and quotas. In the event of unfair competition or manifest crisis, the E.C. can recommend the introduction of antidumping measures or import restrictions on steel products. In practice, most trade measures in steel are coordinated at the Community level.

Causes of Surplus Capacity

In Europe and throughout the world, 1974 was a boom year for the steel industry. The 1974–1975 oil price shock precipitated the crisis in the European steel industry. Industry analysts, however, now contend that the state of excess capacity would have occurred even without a recession.

Table 2.1 presents annual data on production, capacity, and capacity utilization in the European Community from 1974 to 1986. Production fell by nearly 20 percent from 1974 to 1975, and though it subsequently fluctuated with the business cycles, it never approached its 1974 level. Capacity, moreover, continued to rise until 1980, resulting in low capacity utilization throughout the period.

Several factors accounted for the excess capacity. Consumption of steel within E.C. countries has leveled off, following its rapid growth in the post–World War II period. Important steel-using industries in Europe, such as ship building, have declined. In addition, steel input per unit of output has dropped in a number of important downstream industries. Many products have become lighter. Modifications in prod-

Table 2.1. European Community production and capacity, crude steel, 1974–1986.[a]

Year	Production (in thousands of tons)	Percent of 1974 production	Capacity (in thousands of tons)	Capacity utilization (in percent)
1974	155,587	100.0	178,930	87.0
1975	125,235	80.5	190,053	65.9
1976	134,156	86.2	197,611	67.9
1977	126,121	81.1	199,900	63.1
1978	132,580	85.2	202,119	65.6
1979	140,195	90.1	203,469	68.9
1980	127,738	82.1	202,536	63.1
1981	125,144	80.4	197,885	63.2
1982	110,509	71.0	194,554	56.8
1983	108,668	69.8	186,244	58.3
1984	119,236	76.6	172,851	69.0
1985	119,668	76.9	167,008	71.7
1986	111,987	72.0	161,520	69.3

a. Figures include data only for the original nine countries: West Germany, France, Italy, the Netherlands, Belgium, Luxembourg, Britain, Ireland, and Denmark. Data are taken from EUROSTAT, *Iron and Steel Yearbook,* for the years 1974–1986.

uct design and production processes have increased steel yield by lowering scrap loss. In addition, materials such as plastics, glass, and aluminum have been substituted for steel. (With recent technological developments in special steels, however, steel has replaced other materials for some uses.)

The E.C. steel industry exports about one fourth of its product to nonmember countries, and historically has relied on world markets as a vent for surplus during periods of weak domestic demand. However, export markets also were weak during this period.[3] With the general slump in steel demand in industrialized countries, only demand in Third World countries continued to grow, though at a slower pace. E.C. exports to these markets, which accounted for 43 percent of total exports in 1978, met stiffened competition from two sources. First, Third World countries rapidly developed their own steel industries, and began turning from net importers to net exporters. Second, other exporters from industrial countries, mainly Japan, aggressively expanded sales in these markets.

In the markets of other industrialized countries, E.C. exporters met growing protection. The United States has been the largest single export market for E.C. producers, accounting for 22 percent of total exports in 1978. Protectionist measures—first the antidumping trigger price mechanism and, subsequently, voluntary export restraints on major product lines—effectively blocked expansion of E.C. exports into the United States.[4]

According to an E.C. Commission report, "stagnating steel output in the Community combined with slack internal consumption and an industry not competitive enough to increase its exports would not in themselves have brought about the structural imbalance if at the same time production capacity had not increased substantially since 1974, leading to a serious and persistent gap between supply and demand" (Commission of the European Communities 1983, p. 6). The expansion of capacity was due to a combination of new investment, productivity improvements, and failure to close obsolete plants.

During the 1973–1974 boom, many firms decided to keep marginal plants in production or to restart such plants, and drew up plans for extensive capacity expansion. Most perceived the slump in 1975 as temporary. Not until 1977 did the E.C. Commission begin stressing the structural component of the decline. Although industry assessments varied, recognition by companies that the problem was permanent tended to come later. The steel industry has always been exposed to cyclical fluctuations, and firms tended to be more concerned with positioning themselves favorably to respond to the expected surge in demand.[5]

Productivity improvements also contributed to the problem of excess capacity. As a whole, the European steel industry lagged far behind the most efficient world producers. Insufficient investment in new capital and poor labor relations in certain countries were contributing factors. The drop in domestic steel demand and prices increased pressure on companies to modernize capital and improve labor productivity in order to cut costs and be more competitive in world markets. The improvements were dramatic. The E.C. Commission estimates that between 1972 and 1980 tonnage yields increased an average of 47 percent for blast furnaces, oxygen converters, and arc furnaces; 69 percent for continuous casting; and 105 percent for strip mills (Commission of the European Communities 1983, p. 42). These gains in productivity, however, magnified the problem of surplus capacity, as they were not always accompanied by the closure of inefficient plants.

The Development of a Community Restructuring Plan

Between 1974 and 1975 steel production in the European Community dropped by almost 20 percent. Existing market discipline dissolved, as internal prices plummeted an estimated 40 percent and export prices 50 percent. The leading European producers fell into two opposing groups with respect to their strategies for restoring market order. The German approach was to extend the private cartel organization. In January 1976 the German producers formally expanded the existing organization to include ARBED in Luxembourg, Hoogovens in the Netherlands, and ARBED's Belgian Sidmar group. The new cartel, known as DENELUX, was sanctioned by the E.C. Commission. The French and French-speaking Belgian producers were especially strong opponents of the new cartel.

The French producer's association, whose principal members, Usinor and Sacilor (then privately held), were experiencing some of the largest losses in the European Community, instead pressed for the declaration of a manifest crisis and the establishment of production quotas by the commission. The French were joined by the British and Italians in calling on the E.C. for external protection and other internal stabilization measures.

The Germans, in particular, opposed any intervention on the grounds that the crisis was cyclical, not structural, in nature. Yet one analysis noted that, "as usual, ideology was conveniently married with economic interests. Left on its own, the steel industry was bound to be faced with many bankruptcies and most German firms, because of their relative competitiveness, were likely to ride out the storm" (Tsoukalis and Strauss 1987, p. 197). The German, Dutch, and Luxembourg firms generally were in a strong position to survive price wars, because their financial losses, though substantial, were usually small relative to those of companies in other countries.

Largely because of German opposition, the E.C. Commission maintained that the problem was cyclical and took only limited steps to address the crisis. These measures primarily involved more careful monitoring of the market through increased data collection and forecasting.

As the market situation continued to deteriorate, pressures for E.C. intervention mounted. Although the French and Italian firms continued to be the strongest proponents of E.C. intervention, companies in other countries that at first had opposed all measures began to reverse their position. The initial strategy of the German, Dutch, and Luxembourg producers was to win victory through attrition. That strategy failed, however, when governments in other countries intervened with massive production subsidies to avert bankruptcies. The withdrawal of capacity by the least competitive firms would have entailed massive layoffs in areas where unemployment was already high and where regional economies were highly dependent on steel. Britain, France, Italy, and finally Belgium began heavily subsidizing their industries to prevent plant closures. Without substantial capacity reduction, however, the financial viability of even the strongest German companies was undermined.

Government subsidies often were channeled to publicly owned steel companies. Public ownership of the steel industry grew significantly during the crisis, particularly in France and Belgium. Table 2.2 presents comparisons across E.C. countries of the percentage of the steel industry owned by the government in 1965 and 1981. Not surprisingly, countries with the least government ownership (Germany, the Netherlands, and Luxembourg) tended to be the countries with the least government subsidization.

Accompanying the government subsidization were threats by certain countries to restrict intra-E.C. steel trade, also illegal under E.C. law. Producers, particularly in Germany, France, and Italy, pressed for protection against imports from other E.C. countries. Governments responded with threats and certain limited actions. A consensus soon developed, however, that all would be net losers were the common market in steel ended: not only would there be disruptive effects on the steel market itself, but also the conflict could spill over into other

Table 2.2. Government ownership of the steel industry, 1965 and 1981 (in percent).[a]

Country	1965	1981
Belgium	0	57
France	0	70
West Germany	10	11
Italy	60	60
Luxembourg	0	0
Netherlands	0	36
U.K.	8	76

a. Data are taken from Hogan (1983).

markets. As one analysis concluded, "the most important thing that united the Community countries during the crisis was the need to preserve the common market and thus avoid the spread of national protectionist measures" (Tsoukalis and Strauss 1987, p. 216). Therefore, producers with a large internal market, such as the Germans, could not rely on trade protection from imports originating from other E.C. countries, and it was in their interest to cooperate in an E.C. restructuring plan.

In 1977 the European Community finally declared that the crisis was structural in nature and implemented limited measures. Over the next few years, these measures were strengthened and expanded and became known, collectively, as the Davignon Plan, named after the new E.C. commissioner for industrial affairs. The Davignon Plan had two facets: short-run measures designed to improve prices and profitability, and long-term measures for restructuring the industry in accordance with the new levels of demand.

Short-Term Measures

The initial short-term measures included (1) mandatory minimum prices for selected categories of products; (2) guide prices for other steel products; and (3) voluntary restrictions on production for a number of products not covered by minimum prices. To negotiate and enforce production restrictions on a company-by-company basis, a new cartel was created to include all major E.C. producers. Members of the new organization, EUROFER, accounted for about 85 percent of E.C. steel production. The E.C. Commission and the producers set product-specific quotas. The E.C. Commission determined quota rights for companies using a complex formula based on production levels for 1974, a period when firms presumably were operating at close to full capacity. Firms then traded quota allocations within EUROFER to establish final quotas. In some instances, the mandatory minimum prices helped stabilize prices in markets where small producers, who were not members of EUROFER, had large market shares.

In addition, measures were taken to control disruptive imports. In 1978 the E.C. established antidumping regulations and instigated fifteen antidumping suits; in twelve of these suits, firms were forced to pay antidumping duties. The E.C. signed bilateral, voluntary restraint agreements with the fifteen countries in 1978, and these agreements were renewed annually.

Superficially, the Davignon Plan could be read as a means of controlling imports (through external measures), while sanctioning the cartel organization of the rest of the market (through the creation of EUROFER and production quotas, and through the imposition of mandatory minimum prices). However, a deeper, symbiotic relationship seems to have evolved between EUROFER and the E.C. Commission. The commission played a crucial leadership role in mediating differences between producers. When agreement periodically broke down, the commission intervened and imposed production quotas in an effort to uphold market discipline. The commission, with an ultimate mandate from the member states to preserve the common market, also relied on the producers. Without their cooperation, its measures, which were intended to restore market order, would be unenforceable. The fact that the system relied on producer cooperation was used to justify an approach that strengthened the operation of the cartel.

In 1980, with the second oil price shock, steel demand plummeted and EUROFER dissolved, as the German firm Klöckner and the Italian firm Italsider slashed prices and flagrantly violated production quotas. The E.C. Commission's efforts to restore EUROFER immediately after the breakup failed, primarily because of tensions among the German producers over Klöckner. As a result, Davignon requested that the Council of Ministers declare a state of manifest crisis and allow the commission to exercise its treaty powers to impose production quotas. The German government at first vetoed the proposition, but subsequently reversed its position; it wanted to protect the common market in steel and had become concerned over the growing financial losses of German companies. In October 1980, for the first time in its history, the E.C. declared a state of manifest crisis and substituted mandatory for voluntary production quotas.

In the 1980s the steel industry was governed by a succession of policy regimes designed to stabilize the steel market in the short run. All combined direct imposition of E.C. policy on the producers with voluntary agreements in a reconstituted EUROFER. In essence, when EUROFER producers were unable to reach an agreement on production quotas for certain product lines, the commission intervened and imposed a quota system.

In efforts to raise prices and restore profitability, the succession of E.C. policy regimes exercised tighter and tighter control over production, prices, and deliveries. The series of voluntary agreements and compulsory measures increased the number of product categories covered by quotas. The commission also tried to influence prices more

summaries of the structure of the industry, problems of competitiveness, and strategies for restructuring are provided for each country.

West Germany

The German steel industry is the largest in Western Europe. The industry contains a number of large, integrated producers, including Thyssen, Krupp, Klöckner, Salzgitter, Mannesmann, and Hoesch. Of these, only Salzgitter was, until recently, government owned. In 1978 ARBED of Luxembourg purchased all of the steel capacity in the Saar, forming ARBED-Saarstahl.

Unlike most in the European industry, German steelmakers had a well coordinated, long-run strategic plan for improving productivity in the industry. Since the mid-1960s, German producers, through various joint sales agencies and rationalization groups, had shared orders to improve productivity and had coordinated the introduction of new capacity. In addition, the German companies tended to be more diversified than other European producers at the beginning of the crisis. Diversification into nonsteel industries or "upmarketing" product lines within the steel segment accelerated during the crisis.

Initially, the German government had little involvement in the restructuring of the steel industry. A policy of nonintervention generally is credited to the government's free-market philosophy. Subsidization of the steel industry in the Saar was an important exception, however, and indicates that in the face of severe economic disruption, the German government would show little reluctance to intervene.

As a result of changes in optimal plant location, the Saar had declined as a steel-producing area and had experienced particularly serious structural adjustment problems. The Saar is the poorest state in West Germany, and its economy is highly dependent on steel. Originally located in the Saar because of the region's abundance of raw materials, the industry was now handicapped by the increasing scarcity and cost of these materials, and by its distance from German markets. Furthermore, its plants were small and antiquated. In 1978 the German government began providing the Saar steel industry with large infusions of funds so that it could modernize its plants. From 1978 to 1986 the federal and Saar governments in Germany provided ARBED-Saarstahl with about $1.4 billion in aid. In 1986 the German government acquired effective control of the company.[10]

As the crisis continued, the German government became increasingly involved, more generally, in planning and financing the long-term restructuring of its steel industry, while pressing for a halt to all state subsidies at the Community level. The government promoted rationalization through merger. However, numerous attempts to merge various integrated producers all failed.

France

Although prior to the crisis the French steel industry had been privately owned, the government had exercised considerable control over the industry since the 1940s through planning agreements, provision of investment capital, and price controls. From the beginning of the crisis, the French steel industry suffered large losses, and by the end of 1976 the industry, threatened with bankruptcy, sought state aid. The government of President Giscard d'Estaing provided aid as a stopgap measure, but delayed major initiative on the steel crisis until after the 1978 general election. In September 1978 the Giscard government restructured the finances of the largest steel producers, which accounted for about 80 percent of French steel production, and consolidated them into two companies: Usinor and Sacilor. The government converted much of the steel industry's massive debt ($9 billion, or about 120 percent of a year's turnover) into equity, keeping 15 percent for itself and giving about 60 percent to state-run banks and credit organizations. The Socialist government of President François Mitterrand officially nationalized the two companies in 1981 and merged them in 1988.

At the beginning of the crisis, the French industry was in a weak competitive position because of the age of its plants and the resulting low productivity. The French industry has two modern, coastal plants, Fos-sur-Mer and Dunkerque. Its older, inefficient plants were concentrated in the Lorraine and Nord regions.

Toward the end of 1978, Raymond Barre, the French prime minister, unveiled an ambitious restructuring plan that called for eventual reduction of the labor force from its peak of 155,000 in 1974 to 110,000, and for the closing of a quarter of the steel plants. The Barre Plan announced an immediate program for 22,000 of these cuts, most of which were to be in the Lorraine. The announcement sparked the worst rioting in France since May 1968, with violent protests in the Lorraine itself and a march of 100,000 steelworkers and sympathizers in Paris. The government subsequently modified its plan, agreeing not to close any plants before new jobs were available. The Barre Plan, however, did achieve substantial capacity reductions and productivity improvements.

Despite gains, French steel remained in financial trouble and constituted a serious drain on state resources. In 1983 steel siphoned off one third of the total budget for nationalized industries (about $1 billion out of $3 billion). In the debate over industrial policy, funding of declining industries such as steel became increasingly controversial. French industrial policy was intended primarily to support growth industries such as electronics, nuclear power, arms, and biotechnology. Support of declining industries encroached seriously on funding for high technology, hindering the effectiveness of these programs.

In 1982 the Mitterrand government announced a new steel plan calling for extensive plant closings in France's older steel regions, and further concentration of production at its newer coastal sites. Once again, the announcement precipitated rioting and marches on Paris. The government revised its steel plan in 1984, and reached an accord with the more moderate French unions that paved the way for further consolidation in the industry.

Belgium

The deep schism in Belgian society between French-speaking and Flemish-speaking populations compounded the social difficulties of restructuring, which necessarily involved large cuts in capacity and employment. The Belgian industry consists of several large integrated producers and a few specialty producers. The large integrated producers are Cockerill-Sambre, Sidmar, Boel, and Clabecq. Sidmar, owned by ARBED of Luxembourg, comprises Belgium's most modern works; it is located on the coast in the Flemish-speaking region of Flanders. Cockerill-Sambre is Belgium's largest company, and has experienced the greatest financial difficulties; its plants are located in the economically depressed French-speaking region of Wallonia.

Cockerill-Sambre was formed by the reorganization and merger of Belgium's two largest steel producers, Cockerill and Hainaut-Sambre, which had been on the verge of bankruptcy. The government acquired more than 50 percent of the companies. Cockerill-Sambre's financial difficulties stemmed from inefficient plants and poor industrial relations. With respect to the former, the company had some of the most advanced rolling equipment in Europe, but suffered from old, poorly located, and badly laid out plants in basic steel. Controversy surrounding government proposals to slash capacity at Cockerill-Sambre led to its fall in 1981. Government restructuring plans subsequently focused on productivity improvements, with no clear plan for capacity reductions.

In an effort to break the stalemate, the Belgian government hired

Jean Gandois, a French businessman and former head of the French steel company Sacilor, to make recommendations for restructuring. The main elements of the Gandois Report, presented in January 1983, called for the development of links between Cockerill and Hainaut-Sambre and the closing of two out of the five plants. Gandois subsequently assumed the presidency of a united Cockerill-Sambre and carried out restructuring under a modified version of this plan.

Italy

The Italian steel industry comprises Italsider, which is a large, integrated, government-owned producer, and numerous electric furnace companies, which are small, privately owned operations. Italsider has three fully integrated plants. A concentration of about eighty small plants in northern Italy, known as the Bresciani, primarily produce rods. Collectively, the Bresciani have a substantial effect on this segment of the market. The Bresciani are efficient and renowned for their flexibility in production.

The Italian government has used the steel industry as a tool of economic development, especially in the south. Italian production grew rapidly from 2.4 million tons in 1950 to 24 million in 1978, paralleling a similar growth in demand. After 1975 domestic demand stagnated, and the scope for exports narrowed. In the first half of the 1980s, Italsider sustained record losses averaging $1 billion per year. Depreciation and debt servicing accounted for about 80 percent of these losses. Italsider acquired much of its debt in the early 1970s to finance expansion.

In the early years of the crisis, the government attempted to cut costs through modernization programs; it was especially resistant to any capacity cuts. Initially, the E.C. Commission justified Italy's position by the fact that its domestic demand was stronger than that of other countries. It also cited equity considerations: Italy's economy was still relatively undeveloped. As the crisis worsened, pressure mounted from other E.C. countries for Italy to bear its share of the burden of restructuring. As a result, in June 1983 the commission ordered Italy to reduce its capacity by 20 percent within two years. Eventually, Italy worked out a restructuring plan with the commission; a main feature was the closing of a major facility of the state-owned company, Italsider.

The Netherlands

The Netherlands has one large integrated steel producer, Hoogovens. The national government and local government have a minority interest in the company. Government involvement in managing the company, however, has been minimal.

Hoogovens' works is one of the most modern in Europe. Located on the coast, it has a transportation cost advantage in importing cheap iron ore and exporting finished products. As with all of the major E.C. producers, however, Hoogovens has absorbed capacity cuts as part of the E.C. restructuring plan.

Luxembourg

Steel has dominated Luxembourg's economy. Commenting on the size of the country relative to the E.C. steel industry, William Diebold wrote, "On paper Luxembourg looks like a group of steel mills with a line drawn around them . . . Statistically, the effect is rather like that of establishing an international frontier around several counties in Western Pennsylvania or a shire or two in the Black Country of England" (Diebold 1959, p. 128). Prior to the crisis, iron and steel accounted for about 70 percent of Luxembourg's exports and two thirds of its industrial employment, or one sixth of total employment. ARBED, which also operates plants in Belgium and, until recently, Germany, is the sole producer of steel in the country.

Luxembourg's extreme dependence on steel necessitated a quick and decisive response to the steel crisis. A tripartite committee of government, labor, and management was formed and coordinated plant closures, employment adjustment issues, and modernization.

Britain

Britain nationalized more than 90 percent of its steel industry in 1967, with the goal of modernizing and restructuring. That task was delayed as a result of controversy surrounding the nationalization itself and the commercial and social objectives of restructuring. Consequently, rationalization was not accomplished in the late 1960s and early 1970s, when growth in demand would have eased the transition. Instead, the British Steel Corporation (BSC) implemented a massive restructuring program entirely during the slump.

In the beginning of the crisis, the BSC was uncompetitive in terms of its costs, product quality, and service. As a result of low levels of investment, much of the BSC's capital was antiquated. Its labor productivity was the worst in the European Community, not only because its plants were old but also because labor relations were poor.

On the eve of the 1974 recession, the BSC launched a major investment program to modernize and restructure the industry. The plan called for an increase in capacity from 27 million tons in 1975 to 35 million tons by the mid-1980s. Production was to be consolidated in five integrated, coastal works in order to take advantage of economies of scale in production and low shipping costs. The downturn in 1975 delayed scheduled plant closures, because the British government kept the plants open to bolster employment.

With losses mounting, the new Conservative government was determined to restore profitability to the industry and return it to the private sector. The Thatcher government implemented the most radical restructuring program in the European Community, abandoning ambitious plans to expand capacity and slashing it instead. Concentrating production at modern, coastal facilities remained the basic restructuring strategy. The strategy magnified the effects of capacity reduction on employment, for it entailed not only a net loss of jobs but also the relocation of jobs away from certain traditional steel-making areas.

At the same time, the Thatcher government moved toward privatizing the industry by selling off assets to private firms. In 1988 the government completed the privatization process with the public sale of the BSC's stock.

Dismantling the Davignon Plan

With the combined results of restructuring in the member states, the European Community's target for capacity reduction under the Davignon Plan was reached, and the E.C. Commission began liberalizing the steel market in 1986. However, because capacity utilization, though improved, remained moderately low at about 70 percent, the system of quotas was to be phased out over a period of several years to allow companies time to complete restructuring.

At that time a new code on aids was adopted, which, though falling short of prohibiting state aid entirely, strictly limited its use. The main exception to the prohibition on state subsidies concerned plant closures. States could reimburse the enterprise for 50 percent of the social costs associated with closure. If the company chose to withdraw completely from the steel business, the state could pay the company for part of its loss of assets. Thus, the new code on aids was designed to remove barriers to the closure of additional capacity from the industry and to minimize the competitive advantage that companies gained from government subsidies.

By 1987, the E.C. steel industry again began to encounter serious problems of excess capacity. This time the problem was almost entirely trade related. While internal demand remained fairly stable, exports dropped sharply. This was primarily due to a reduction in imports to members of the Organization of Petroleum-Exporting Countries (OPEC) and to the appreciation of the European currencies against the dollar, which rendered E.C. steel producers generally less competitive in world markets.

Although producers and the E.C. Commission agreed that a problem existed, there was little consensus as to its magnitude or the appropriate solution. In 1987 the commission estimated excess capacity at another 30 million tons. In a proposal similar to the original Davignon Plan, the commission argued that it could extend the quota system only if this were accompanied by substantial commitments by producers to reduce capacity. Although producers, not surprisingly, supported the continuation of quotas on a number of product lines, they argued that the commission was overestimating the extent of excess capacity in the industry and they refused to make additional commitments to reduce capacity. Consequently, in July 1988 what remained of the system of quotas was dismantled. Restructuring would continue henceforth under a liberalized regime.

3

Job Security and the Methods of Work Force Reduction

A major cause of the conflicts over subsidies and trade within the European Community has been the dislocation that would arise from plant closures and massive layoffs in the steel regions. Steel restructuring has been a volatile political and social issue in all countries. Efforts to consolidate and reduce capacity in the industry led to the fall of the Belgian government in 1981 and to the walkout of the Communists from their coalition with the Socialists in France in 1984. The announcement of plant closures precipitated rioting in some of the countries.

The European Community has always recognized that if the common market is to be politically viable, it must have social policies that ease the costs of adjustment to workers and their communities when an industry experiences a sudden and substantial decline in production. Otherwise, national governments, particularly in countries with large domestic markets, would come under tremendous pressure to erect trade barriers against exports from other member states. Reflecting this view, a recent E.C. Commission report outlining policies for the steel industry states, "The Commission insists that all the proposals in this communication form an indivisible whole: the industrial aspects (restructuring and market controls through the quota system) cannot be separated from the social and regional aspects stemming from the restructuring process" (Commission of the European Communities 1987b, p. 1). Work force reduction programs, in practice, have been an integral part of restructuring plans forged at all levels—company, state, and European Community.

Huge job losses have accompanied capacity reductions and modernization during restructuring in virtually all countries, as shown in Table 3.1. Employment in the E.C. steel industry as a whole dropped by 50 percent from 1974 to 1986. The largest reductions occurred in Britain, where the work force shrank by more than 70 percent during this period.

Table 3.1. Employment in the iron and steel industry, 1974–1986.[a]

Year	W. Germany	France	Italy	Nether-lands	Belgium	Luxem-bourg	U.K.	E.C.[b]
1974	232,037	157,833	95,656	25,077	63,738	23,503	194,347	792,191
1975	221,853	155,775	96,140	25,401	59,348	21,447	184,385	764,349
1976	219,142	153,948	98,015	25,066	57,198	21,755	182,298	757,422
1977	209,465	142,992	96,593	23,293	49,752	17,437	178,874	718,406
1978	202,801	131,595	95,591	21,295	48,541	16,774	165,361	681,958
1979	204,813	120,555	98,720	20,931	48,665	16,351	156,396	666,431
1980	197,406	104,940	99,528	21,047	45,220	14,904	112,120	595,165
1981	186,685	97,305	95,651	20,911	44,106	13,419	88,247	546,324
1982	175,946	95,200	91,495	20,158	41,649	12,425	74,475	511,348
1983	163,748	90,714	87,050	19,210	39,569	12,934	63,694	476,919
1984	152,467	85,064	75,611	18,748	37,184	12,713	61,856	443,643
1985	150,833	76,141	67,408	18,780	34,542	12,612	59,056	419,372
1986	142,713	68,404	65,582	18,933	30,535	12,274	55,872	394,313
Percent change, 1974–1986	−38.5	−56.7	−31.4	−24.5	−52.1	−47.8	−71.3	−50.2

a. Numbers represent end-of-year employment of manual and salaried workers. Data are taken from EUROSTAT, *Iron and Steel Yearbook,* for the years 1974–1986.

b. Includes workers only in these seven countries.

Apart from the sheer magnitude of the job loss, several factors have complicated the development of work force reduction measures. The onset of the crisis in steel coincided with a curtailment of overall growth in most European economies. Compounding this problem was the fact that the steel industry often dominated regional economies. Consequently, most job vacancies have been located in different parts of the country, and steelworkers, who often have strong ties to the region, have been reluctant to move.

The onset of the crisis also coincided with a Community initiative and legislation in many E.C. countries that increased protection of workers in the event of layoffs. This legislation, which typically requires companies to use alternatives to layoffs and which increases the participation of worker representatives in company decisions concerning layoffs, substantially increased workers' rights in jobs.

Moreover, traditional company personnel practices raised workers' expectations of job security. Many companies had provided generations of workers with stable employment, and workers had come to expect job security. Though any job rights stemming from historical company practices were merely implied, they were arguably no less important in shaping work force reduction programs than those established by law.

The European Community and member states were caught in a dilemma in which, on the one hand, they were pressing for rapid restructuring and, on the other, they were promoting full employment and strong job rights. Similarly, companies had to balance the need to reduce capacity and increase labor productivity against the assurances of stable employment that they traditionally gave to workers. Governments often participated in labor negotiations, and they along with the European Community always provided crucial financial backing for the measures adopted. Ultimately, however, work force reduction measures were the product of negotiations between companies and unions. The methods of work force reduction varied considerably across countries; so too did the level of job security for steelworkers, and the adjustment costs borne by workers, companies, and the public sector.

Labor Mobility and Adjustment Problems in Steel Regions

Problems of economic dislocation are not new to the European Community, and past experiences have guided recent policies with respect to steel. One of the first major tests of the European Coal and Steel

Community concerned work force reductions in the Centre-Midi region of France. An estimated 5,000 miners were to be laid off between 1953 and 1956, but the government believed they could be absorbed in the coal mines of Lorraine. Thus, the French government, in conjunction with the ECSC, established a program to resettle workers. Participation was voluntary. Workers who agreed to be transferred were promised an indemnity to cover lost work time, a moving allowance, housing in Lorraine, higher wages, and more hours. Yet only one-tenth of the expected number volunteered, and in 1956 the program was abandoned.

A bishop writing to the head of one of the Catholic unions, which opposed the transfer, stressed the importance of social factors in the program's failure: "You are right . . . when you say that you are attached to your soil, to your country, to the cemetery where you are buried, to your living relatives . . . I can only approve the fact that you feel these things, which some, who dream of nothing but machines and material things, seem unable to feel any longer. You, on the contrary, grasp the higher quality of these emotions which have their source and their echo in your heart" (quoted in Diebold 1959, pp. 408–409). In the wake of the abortive program to relocate French coal miners, the ECSC sponsored several studies to evaluate the entire readjustment assistance process. One study of the workers who did transfer concluded that "the transferred workers assimilate themselves to the local population only with great difficulty. In practice there has been no, or hardly any, assimilation up to the present" (quoted in Diebold 1959, p. 409).

The attitudes and policy issues of today are little different from those of some thirty years ago. Each individual interviewed for this study, without exception, stressed the workers' attachments to communities and their opposition to moving. Based on extensive interviews with steelworkers who had been laid off in Wales, Harris (1984) makes similar observations. Harris stresses the importance of social networks, particularly contacts with kin, for support following layoffs, and points to the strong attachments steelworkers had to their family and communities to explain low migration from steel regions, despite high unemployment in those areas. Often generations within families had worked for the same company. In practice, little use was made of relocation schemes.

Summarizing the situation of European steelworkers, William Sirs, president of the Iron and Steel Trades Confederation in Britain, remarked, "There is a general reluctance to move out of traditional steel regions, even if the move is to another such region, especially if workers from other deprived areas have been integrated into a steel community . . . There are some incentives to move, but it is hardly likely that they will overcome the natural reluctance to move, and in any case the areas of full employment are fewer and fewer" (Sirs 1980, p. 46).

Traditional company policies often reinforced steelworkers' natural reluctance to move. Examples from the German steel industry illustrate this interrelationship between company policy and worker mobility. Historically, German firms monopolized local labor markets. By purchasing huge tracts of land and controlling local politics, they prevented other industries from competing for their labor.[1] Following World War II, when industries were relocating from East to West Germany, many moved to the southern part of West Germany after effectively being blocked by the older, established industries in the north, which included coal and steel. The monostructured economies of the steel regions, then, were not always the work of competitive market forces. On the supply side, companies have discouraged worker mobility by providing employees with company housing and bonuses for long-term attachment. Companies often have employed several generations of workers from a particular family. In light of these practices, the German unions have vehemently argued that the steel companies are obligated to provide their workers with jobs.[2]

Available evidence concerning the steel regions in Europe clearly shows the structural problems these areas are facing. In a period of high and growing unemployment in Europe, unemployment rates in the steel regions generally have been well above national averages (Table 3.2). The rapid decline of employment in the steel industries has resulted in considerable structural unemployment. Migration often is viewed as a solution to unemployment in declining regions. However, it can have serious adverse economic consequences. Although there has been migration from these regions, according to the European Commission, those who have relocated have been mostly young people. Such migration, in the words of a commission report, "is likely to weaken the social fabric of the areas" and may make the region a less attractive place for new investment (Commission of the European Communities 1987b, p. 7).

Together, economic and social factors—the high regional dependence on steel for employment, general economic stagnation in the

Table 3.2. Unemployment in steel regions, 1979 and 1981 (in percent).[a]

	1979		1981	
	Region	Country	Region	Country
West Germany	4.3	3.0	4.7	3.6
Belgium	9.1	7.3	14.2	12.0
France	7.3	6.6	8.7	7.8
Italy	6.3	6.7	7.3	7.6
Netherlands	7.7	6.5	10.8	9.0
Luxembourg	2.4	—	2.0	—
U.K.	7.3	5.3	13.2	9.8

a. Regional figures are weighted averages of steel regions where the weights are employment in steel. The region corresponds to the level III region in the E.C. classification system. Unemployment data are from EUROSTAT. Data on employment in steel by region were calculated by the author from unpublished, E.C. plant-level data. For Luxembourg, the country corresponds to the region.

regions, labor immobility, and the tradition and expectation of strong job security for workers—have complicated work force reductions and have encouraged parties to seek alternatives to layoffs.

Job Security and Social Policy: The Role of the Community and Member State Governments

Social policies of the European Community and member state governments have affected work force reduction in steel through two channels. First, labor legislation has broadly defined worker rights in the event of layoffs and, arguably, has increased union bargaining power. Second, and in a more direct way, general welfare programs and special programs targeting the steel industry have helped finance the costs of income or job security for workers.

European workers have always had some legal rights in the event of layoffs. In general, European law has afforded individual workers basic employment protection since the nineteenth and early twentieth centuries. For example, in Germany, laws protecting workers against unjustified dismissal and requiring advance notice prior to dismissal date back to the 1920s. Although Britain did not pass legislation requiring advance notice until 1963, British common law had required such notification since the mid-nineteenth century.[3]

Restructuring in the steel industry has taken place during a period when many Community countries were substantially increasing the rights of workers in the event of layoffs. In the late 1960s and 1970s, with the exception of Italy, all E.C. countries in this study enacted legislation governing collective dismissal for economic reasons. Much of the legislation was a response to a 1975 E.C. directive requiring member states to provide minimum levels of protection against layoffs.

Obligations to employees typically include a minimum notification period, a minimum severance payment, and a requirement to inform and consult employee representatives. The strength of all of these provisions varies across countries, but differences in consultation requirements are perhaps the most subtle and most important. For example, in Britain, the unions have only the right to present alternatives to layoffs which management must consider. In Germany, where the provision is the strongest, management must negotiate with the works council a social plan that stipulates benefits for workers who are laid off. Although Italy does not have collective dismissal laws, industrial workers are covered by a 1965 agreement with the employers' confederation that contains provisions similar to those in the laws in other E.C. countries.[4]

During the 1970s and early 1980s, countries generally strengthened job security legislation. The exception is Britain, where the Thatcher government weakened existing provisions in 1980. More recently, some Continental countries, notably Germany, have weakened job security legislation. The modifications mostly have involved granting exemptions from the requirements of the laws, and have not altered basic rights to notice and participation.

The job security obtained by steelworkers in Community countries generally far exceeded that guaranteed by law, and it would be incorrect to conclude that the work force reduction measures were imposed by law. However, laws may affect the process by which work force reduction programs are determined. Unions in all countries (albeit to varying degrees) were partners in the formulation of work force reduction policies in steel. By setting minimum standards in the event of collective dismissal and by establishing a role for worker representatives in the decision-making process, the law helps legitimize workers' claims in their jobs.

In addition to promoting job security through legislation, the Community and its member states have helped underwrite the costs of income and job security with a variety of programs. The Treaty of Paris stipulates that member states provide steelworkers with certain minimum benefits in the event of layoffs. These benefits include reimbursement for moving costs, retraining benefits, and some guarantee of income in the event that laid-off workers find new employment. The European Community refunds member states for 50 percent of their expenses under certain programs for steelworkers. Precisely what expenses are reimbursable is determined by the Treaty of Paris and by bilateral agreements between the European Commission and the member state. E.C. funds to support social measures for steelworkers come primarily from a tax levied on steel companies. However, during restructuring, these funds were supplemented from the E.C.'s general revenues.

Originally, most of the European Community's refunds covered vocational training, resettlement, and supplementary income programs. In light of the difficulty steelworkers had in finding new employment, the E.C. Commission and the E.C. Parliament adopted measures advocating early retirement and reorganization of working time through short-time work (the temporary reduction of the work week below normal working hours), the introduction of an extra shift, and restrictions on overtime to avoid layoffs. They have backed these declarations with extensive funding of early retirement and short-time work schemes. Social assistance for steelworkers has been substantial. The commission estimates that the European Community assisted two-thirds of the steelworkers who lost their jobs between 1976 and 1983. In 1985, E.C. aid for steelworkers amounted to 141.8 million ECUs.[5]

In addition to aiding steelworkers who are directly affected by work force reductions, the Community has targeted the steel regions for special economic development assistance. The measures vary from traditional low-interest loan and grant programs to the establishment of so-called business and innovation centers.

Social and regional policy, however, is largely the jurisdiction of the individual member states. The growing support by the Community for short-time work, early retirement, and similar measures mirrored policies, especially in the Continental countries, to promote alternatives to layoffs in a period of high and growing unemployment.

Methods of Work Force Reduction

Although job losses have been substantial in all the countries in this study, the methods of work force reduction, the level of job security for workers, and the degree and form of public sector support have varied considerably across countries. The appendix to this book summarizes the policies on work force reduction for each country in four categories: early retirement; transfers; short-time work and reduction of the work week; and benefits to laid-off workers and incentives for voluntary departure. Whenever available, benefits levels and sources of funding are reported. Since the work force reduction measures changed over time in most countries, the information in the appendix reflects their status in 1984, unless otherwise indicated. The discussion that follows highlights the principal methods of work force reduction over the period for each country covered in the study.

West Germany

The job security of German steelworkers reflects, in part, the relatively strong position of German labor. Under German law, companies must inform and consult labor representatives in the event of layoffs, and negotiate with them to develop a social plan. The process occurs in three stages. First, management must consult with the works council about alternatives to layoffs or plant closure. Next, the supervisory board must approve the layoffs or closure. If these are approved, management must then negotiate with the works council to develop a social plan, designed to alleviate the economic effects of the layoffs.

In addition to a system of works councils at the establishment level, German law requires labor representation on the supervisory boards in large companies. This system of codetermination originated, and remains the strongest, in the steel industry. With equal representation on the supervisory board, labor has been able to block closures and negotiate favorable terms in social plans. According to the iron and steel employers' association (the Arbeitgeberverbandes Eisen- und Stahlindustrie), compensation levels have varied substantially across works, but the framework of the agreement has not. The government has not been a direct party to these agreements, in contrast to the situation in other countries.

The methods of work force reduction in the German steel industry are representative of those used in a number of Continental countries. The German steel industry accomplished work force reductions with minimal recourse to layoffs. Instead, companies relied extensively on early retirement, the reduction of hours worked, the transfer of younger workers into positions vacated by retiring workers, and job buyouts.

Early retirement required mutual consent of the employer and the employee. The age of early retirement in most companies was fifty-five; in the Saar it was fifty. The early-retirement scheme drew upon state unemployment insurance funding, and therefore many retirees officially were fired. They then received unemployment insurance, which was supplemented by the company according to the social plan.

German companies have offered a severance payment to those who voluntarily leave their jobs, a policy designed to accelerate attrition rates. In addition, in light of high unemployment, during 1983 and 1984 the German government offered foreign workers a large lump-sum payment if they returned with their families to their home country. As a result, the job buyouts were particularly successful in luring away foreign workers. During the year the offer was available, 3,700 steelworkers, primarily Turks, accepted.

Various schemes to reduce working time have helped spread the amount of available work over the number of existing employees. German companies have used short-time work extensively throughout the crisis. In 1984, for example, in the German steel industry as a whole, 27 percent of the paid hours were not worked. Workers received benefits from the state social insurance scheme for the shortfall in hours. Steel companies typically supplemented this benefit.

In addition to temporary short-time work schemes, a permanent reduction in the work week has been implemented twice. Currently, the work week in steel is 36.5 hours, which is the lowest in German industry.[6]

France

Prior to 1985, work force reduction measures in the French steel industry were similar to those in Germany and other Continental countries. The key methods were attrition, early retirement, job buyouts, and hours reduction. Since 1985 these measures have been supplemented by a retraining and job placement program, called "Contrats de Formation–Conversion" (CFC).

The cornerstone of the work force reduction program has been early retirement. Initiated in 1976, the age of early retirement was progressively lowered such that steelworkers effectively were leaving at age fifty.[7] Companies guaranteed that early retirees would receive 70 to 80 percent of their former wage. The government has provided companies with a subsidy covering about 90 percent of these costs.

In 1978 French steel companies initiated a job buyout program in which workers who voluntarily left their jobs received a 50,000-franc severance payment. Foreign workers who returned to their country of origin also received a 10,000-franc payment and money for travel expenses from the state. An extremely popular program, it also proved extremely costly to companies and was discontinued. More than 3,000 steelworkers left their jobs under this program.

The steel industry developed an elaborate system of transfers to accompany its early retirement and job buyout measures. Workers could be transferred within a works, to another works in the same company, and even to another works in another steel company. A national collective agreement spelled out basic protection for transferred workers. Workers had the right to reject one transfer (that is, had the right to two transfer offers) before they could be laid off. Transferred workers received some protection of seniority and salary; the salary and seniority of older workers were fully protected. Although, where possible, the company had to offer workers jobs in the same region, many workers were relocated to other regions.

The French steel industry also used temporary work-sharing schemes and more permanent reductions in the work week, in order to absorb excess labor. Workers on temporary short-time arrangements received unemployment compensation for the hours not worked. The French government lowered the work week to 39 hours throughout the economy in 1982. The work week was reduced further in the steel industry, though it varied from plant to plant. Certain plants operated on a 35-hour work week.

With much of the older work force already retired, age measures proved insufficient to cope with additional work force reductions required by restructuring. A 1984 agreement negotiated between the steel employers' association and unions, with the help of the government, introduced an innovative program to move excess steelworkers into other sectors. Although the new program significantly weakened steelworkers' job rights, the provisions were nonetheless generous. Steelworkers who were placed in the program retained formal employment ties with the steel company and were guaranteed 70 percent of their previous salary for up to two years. Steel companies played an integral role in counseling, retraining, and finding new jobs for these workers. To retain an incentive for workers to find new employment on their own, workers could capitalize 65 percent of their remaining benefits if they started a new job or opened their own business.

Belgium

During restructuring in Belgium, layoffs were used in only three instances: twice when enterprises ceased operation entirely, and once when an enterprise closed all of its hot-rolling facilities. In these cases, laid-off workers received special financial assistance as a result of negotiations with the government.

Otherwise, Belgian steel companies avoided layoffs. The principal mechanism of work force reduction was early retirement, which was introduced in 1977. The age of retirement was lowered to fifty-five, and in some cases to fifty. The government helped finance the costs of early retirement, and workers contributed indirectly through wage cuts. A tripartite committee of employer, union, and government officials negotiated the number of reductions. Between 1980 and 1984, 6,746 steelworkers left on early retirement, which represented 84 percent of the net reductions in steel (Groupement de la Sidérugie 1985).

In conjunction with early retirement, Belgian companies employed various measures to reduce working time. The average work week was reduced progressively from 40 to 37 hours for the whole of industry, and to 35 hours at Cockerill-Sambre. In addition, short-time work was used periodically. For blue-collar workers, the government paid unemployment insurance for hours not worked; steel companies usually supplemented this payment.

Transfers within the same company have been common. Transferred workers were guaranteed their income for a specified period of time. Although workers may have been asked to change sites, transfers generally were within the same region.

The Netherlands

The Dutch steel company Hoogovens has not laid off workers during restructuring, relying instead on early retirement and measures to reduce working time. The age of early retirement is sixty; on several occasions, the company lowered it to fifty-seven and a half. The state helped finance early retirement through the social security system, and workers also contributed through a special deduction from their paycheck. Workers transferred were guaranteed their old salary and salary scale, although shift premiums gradually were reduced if those for the new job were lower.

Although during most of the period the average work week remained unchanged at 40 hours, workers gained additional vacation days. The company had considerable discretion in scheduling that vacation time during slack periods. A 1985 agreement lowered the average annual work week for shift workers to 36 hours. This was achieved by implementing five shifts for 37 weeks and four shifts for the remaining weeks. Other workers received thirteen additional vacation days.

Short-time work was used periodically and, as in other countries, was covered by the state unemployment insurance system.

Italy

In contrast to the other European steel industries, the Italian industry continued to expand employment until 1980. Only in the 1980s did the Italian steel industry have to contend with the problem of work force reduction.[8] Under pressure from other E.C. countries to absorb their share of the cutbacks, Italy began trimming its capacity and work force.

The elements of its work force reduction program were similar to those in other Continental countries. The Italian steel industry offered voluntary early retirement at age fifty-five; in 1984, legislation lowered the age to fifty. Workers transferred within the company were guaranteed their former salary for at least eighteen months. To minimize resistance to transfers out of the steel industry, steelworkers reemployed in public works projects received a salary that was 25 percent higher than the public sector salary for a period of eighteen months.

Finally, short-time work, supported by the state social security fund, Cassa Integrazione Guadagni (CIG), was used extensively. In a peculiar feature of the CIG, workers may work "zero" hours. Although effectively not working, they retain a formal employment tie to the company and receive a high percentage of their former pay from the CIG (80 percent of gross income). Because the cost to companies of using the CIG is minimal and because payments to workers from the CIG usually are more generous than those from the regular unemployment insurance fund, the CIG has been a popular mechanism for avoiding layoffs.

Luxembourg

Like other companies on the Continent, ARBED of Luxembourg effected large work force reductions without layoffs. Its approach, however, was fundamentally different. Like other companies, it used early retirement schemes and severance payments to voluntary leavers, in order to accelerate the natural attrition of the work force. Yet management as well as the unions opposed short-time work on the grounds that workers would grow accustomed to being paid to do nothing, and, consequently, that it was bad for productivity.

Instead, ARBED established what it called an Anticrisis Division (in French, "Division Anticrise," or DAC), a separate profit-making center within the company, in which it placed extra production workers. The DAC served as a clearinghouse for transferring workers between production units in the company. More important, workers in the DAC were used in nonproduction jobs within the company, replacing subcontracted labor, or were subcontracted out to the government or to other firms. Workers in the DAC received their former wage.

Table 3.3, which provides figures on employment in the DAC for the period 1977 to 1985, reveals the heavy use of this division to reduce employment in steel production. In the late 1970s, more than 10 percent of the work force was assigned to the DAC; in the recession of the early 1980s, this figure grew to nearly 20 percent. ARBED was fairly successful in using this excess labor. Typically, less than 2 percent of its work force was classified as being in retraining or idle. In the early 1980s, ARBED subcontracted out between 3 percent and 4 percent of its workers to subsidiaries or other companies. In this period, however, the government absorbed a large portion of its excess labor. In 1982, 1,187 workers, representing 7 percent of the labor force, were employed in government works projects.

ARBED and the government also provided financial incentives for workers to find alternative employment on their own. Severance payments were offered to workers who found a job outside steel, provided it was in a growing industry. The government, with assistance from the European Community, also guaranteed steelworkers who found work outside the industry a high percentage of their previous pay for eighteen months. In addition to these voluntary contract terminations, ARBED terminated subcontracted workers, many of whom were foreign. This resulted in the reduction of about 600 workers.

The age of early retirement was fifty-seven. From 1978 to early 1980, early retirement was compulsory; after that, it was voluntary. As a result of generous benefits levels, participation among blue-collar workers was virtually 100 percent. Among clerical workers it was about 85 percent, and among managers it was about 50 percent.

As in other countries, the government, backed by the European Community, helped finance social measures. A tripartite committee of government, management, and labor worked out the details and funding of the work force reductions. The government and the European Community covered virtually all of the costs of early retirement and of the company's losses in operating the DAC.

The workers, in turn, accepted substantial wage reductions to help finance social measures. For 1983–1984, unions negotiated a nominal

Table 3.3. Assignment of Anticrisis Division workers in Luxembourg.[a]

Year	Total employment	Total in Anticrisis Division	Construction and demolition projects	Subcontracted to other firms	Government works projects	In retraining programs or idle	Other production assignments
1977	22,259	2,360	1,252	411	0	391	306
		(10.6)	(5.6)	(1.8)	—	(1.8)	(1.4)
1978	20,944	2,412	1,418	362	0	129	503
		(11.5)	(6.8)	(1.7)	—	(0.6)	(2.4)
1979	19,952	2,187	1,048	362	0	77	700
		(11.0)	(5.3)	(1.8)	—	(0.4)	(3.5)
1980	19,121	2,118	889	324	139	187	579
		(11.1)	(4.6)	(1.7)	(0.7)	(1.0)	(3.0)
1981	18,194	3,628	1,345	574	896	275	538
		(19.9)	(7.4)	(3.2)	(4.9)	(1.5)	(3.0)
1982	16,848	3,357	898	657	1,187	305	310
		(19.9)	(5.3)	(3.9)	(7.0)	(1.8)	(1.8)
1983	15,001	2,570	631	510	709	374	346
		(17.1)	(4.2)	(3.4)	(4.7)	(2.5)	(2.3)
1984	13,948	1,107	178	259	112	184	374
		(7.9)	(1.3)	(1.9)	(0.8)	(1.3)	(2.7)
1985	14,063	637	84	112	58	166	217
		(4.5)	(0.6)	(0.8)	(0.4)	(1.2)	(1.5)

a. Numbers come from company data and are annual averages. Numbers in parentheses indicate percentage of the company's total employment.

wage reduction of 4.6 to 20 percent, which was partly compensated by additional vacation time. Tied to the increased time off was an arrangement allowing ARBED to bank these vacation days from year to year. ARBED could stock up to twenty-two days per worker over a three-year period. Such flexibility allowed ARBED to avoid overtime or new hires in upturns.

Britain

Job security for British steelworkers has been relatively weak.[9] In contrast to the Continental companies, the British Steel Corporation relied principally on layoffs to reduce its work force. In the 1980s, the BSC did not use early retirement or short-time work to mitigate layoffs to the same extent as did other European companies.

Most of the restructuring occurred under the government of Prime Minister Margaret Thatcher, who was first elected in 1979. Her policies toward steel were part of a strategy to privatize and end subsidies to nationalized industries. Her government's handling of the steel unions, in turn, mirrored a broader policy to break the power of labor in Britain.

The BSC began closing plants and laying off workers in earnest in 1978. The early plant closings took place with little controversy, since these plants were the least efficient and had long been targeted for closure. As the crisis deepened, the BSC dropped its plans for capacity expansion and pressed for further plant closures and productivity improvements through the relaxation of work rules. Within the unions, resistance to layoffs became stronger. The conflict culminated in a thirteen-week strike in early 1980. The strike, coming at a time of low demand, was largely unsuccessful. Immediately following the strike, the BSC negotiated a series of plant closures and productivity improvements with local units. Believing that the national union had no power to halt layoffs, local unions concentrated their efforts on obtaining the largest severance payments possible. These payments, which supplemented statutory severance pay, varied considerably across plants. Some were quite large. For example, in the case of the Shotton plant closure, they amounted to a full year's wages, and in many other instances to six months' wages.

In addition to providing lump-sum severance payments, the BSC supplemented state unemployment insurance. Steelworkers also were eligible for a fifty-two-week retraining program at 100 percent of their former gross pay, plus pension benefits from the company. Older workers could exercise a "pension option," which involved adding company unemployment insurance supplements directly to their pension.

According to the rules of the BSC pension plan, men age fifty-five and over (and women age fifty and over) who were laid off received an immediate, nonactuarily reduced pension. An individual whose job was not threatened could volunteer to leave for the purposes of early retirement or for some other reason. In this case, the individual whose job was terminated could be transferred to the department where the individual volunteering to be laid off worked; but he would have to accept the most junior position—irrespective of his years of service or the years of service of the individual agreeing to leave. By failing to guarantee seniority, as was done in other countries, the BSC discouraged such internal transfers and thus the use of early retirement to save jobs for younger workers.[10]

Some of British Steel's more innovative programs have focused on economic development in the steel regions. The BSC created a separate division called BSC Industries to assist with the reindustrialization of the steel regions. BSC Industries' role largely involved providing information to potential investors and establishing small-enterprise workshops.

Comparison of Work Force Reduction Programs and Job Security

The job security afforded steelworkers in the Continental countries was fundamentally different from that in Britain. During the period considered here, Continental steel companies cut capacity with few layoffs. In marked contrast, Britain's restructuring involved massive layoffs, though often with large severance payments.

On the Continent, steel companies accelerated work force reductions through early retirement and severance payments for voluntary departure. France supplemented these methods with an elaborate retraining program beginning in 1985, and Luxembourg created a separate Anticrisis Division, which subcontracted out excess labor. During the adjustment period, companies reduced the work week, increased vacation time, and used temporary short-time work arrangements to absorb excess labor.

Published data on dismissals and layoffs in steel reflect these cross-country differences. Figure 3.1 shows dismissals and layoffs of manual workers as a percentage of the manual work force for the years 1975

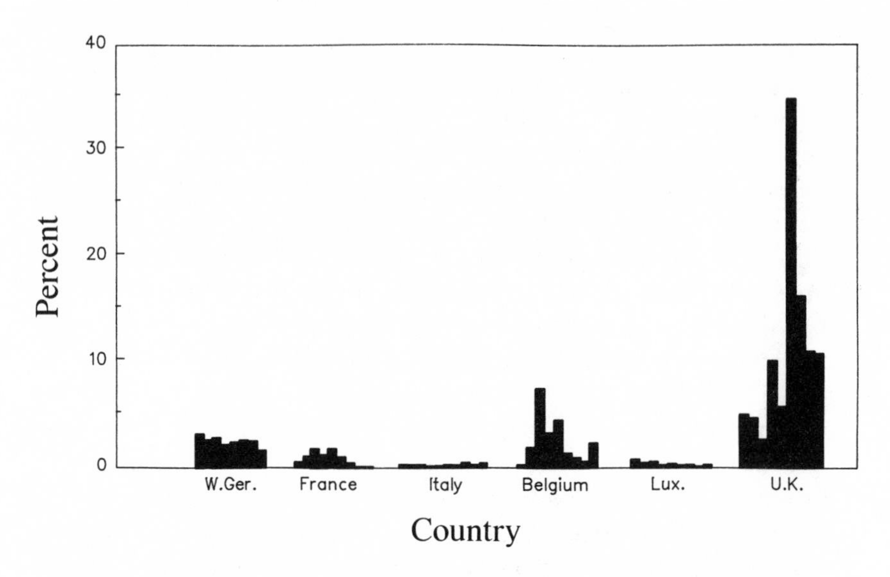

Figure 3.1. Dismissals and layoffs as a percentage of work force, manual workers, 1975–1983. (*Source:* Author's calculations based on EUROSTAT data.)

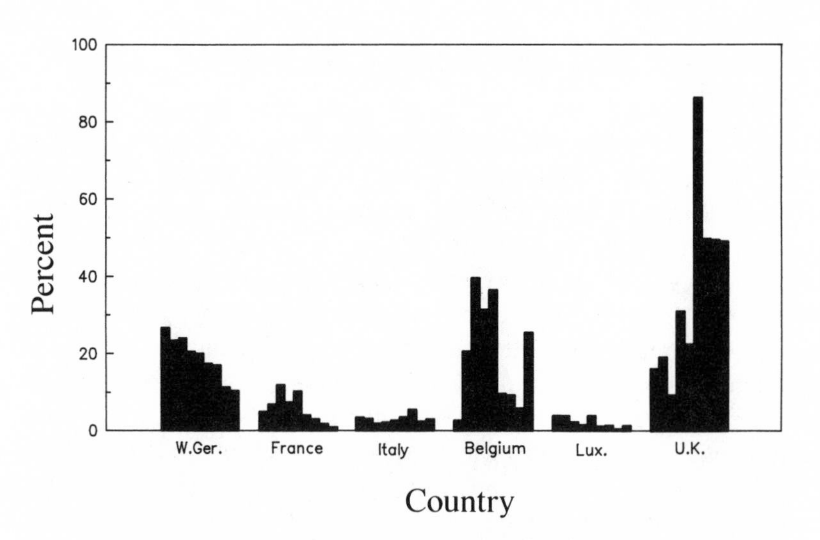

Figure 3.2. Dismissals and layoffs as a percentage of job leavers, manual workers, 1975–1983. (*Source:* Author's calculations based on EUROSTAT data.)

to 1983, for six E.C. countries.[11] Dismissals and layoffs were dramatically higher in Britain than in the Continental countries, particularly in the 1980s.

This difference reflects greater proportionate reductions in the work force in Britain; it also reflects differences in how those reductions were accomplished. Figure 3.2 shows dismissals and layoffs as a percentage of leavers for the manual work force during the same period. Throughout the period, dismissals and layoffs accounted for a negligible portion of leavers in France, Italy, and Luxembourg. In Germany the figures are inflated somewhat by the fact that many workers essentially going on early retirement were officially fired. During the 1980s, dismissals and layoffs as a percentage of leavers were vastly greater in Britain than in the Continental countries. Underlying this difference is the greater use of transfers and early retirement in Continental companies.

The differences in the methods of work force reduction reflected stronger job security in Continental countries than in Britain. The following theoretical and empirical chapters analyze how job security may affect resource allocation in a declining industry and examine such influences on the adjustment of labor and on the allocation of production and capacity during restructuring in the E.C. steel industry.

4

Why Job Rights Affect Resource Allocation: Competing Theories

The most salient fact to emerge from the discussion of work force reduction measures is the widely divergent levels of job security across countries, particularly the differences between Britain and the Continental countries. Differences in the extent of job or income security reflect differences in how the costs of structural adjustment in steel were divided among workers, firms, and the public sector.[1]

Work force reduction programs, with their distributional implications, were the product of negotiations between labor, management, and in some cases government. The outcomes of these negotiations were influenced by a wide variety of factors that helped define explicit or implicit worker rights in their jobs and the relative bargaining power of the parties. These factors include employment protection legislation, government programs, union strength, job security provisions in preexisting collective agreements, company precedent in handling work force reductions, and existing social norms. In this chapter, I do not attempt to formally model the process by which job rights are determined and the costs of adjustment are divided.

Rather, the central question addressed here is: Given the allocation of job rights or adjustment costs between workers and firms, how does that allocation affect employment and, more broadly, restructuring in a declining industry? Intuitively, it would seem plausible that, in the short run, the reduction of employment levels would be less the stronger the workers' rights in their jobs. This outcome, however, would occur only under certain conditions. Explanations based on both neoclassical economic theory and behavioral theories for why job rights may affect resource allocation are examined.

The theoretical arguments draw on the empirical observation that the strong attachments workers have to job and community inhibit labor mobility. I first develop a very simple property rights model, to illustrate the point that work force reductions will be less wherever rights to job security are strong, if workers experience some personal loss (apart from a monetary loss) from being laid off.

I also model the adjustment of employment levels as the outcome of efficient contracts between workers and firms. In this model, the problem is studied directly in terms of how the costs of adjustment are shared; the division of rents (or costs of adjustment) between labor and capital is analogous to the assignment of property rights in jobs. This model highlights the sensitivity of employment, capital investment, and worker compensation in a declining industry to the division of rents. The contract model gives a more extensive treatment of the decision problem than the property rights model, and provides the theoretical underpinnings for the empirical estimates of the effects of job security on resource allocation in the steel industry (presented in Chapters 5 and 6).

The property rights and contract models formally developed in this chapter derive assumptions about human behavior from neoclassical economic theory. The assignment of property rights in jobs or the division of rents between labor and capital affects workers' wealth. The models show that workers essentially will use some of their wealth to secure greater employment stability.

Various behavioral theories, including prospect theory, suggest instead that the assignment of property rights affects individuals' underlying preferences and choices, independent of any wealth effect. Specifically, workers will value their jobs more—or will experience greater disutility from being laid off—the stronger their rights to their jobs. I review empirical evidence of the effect of rights assignments on preferences, and explanations of this phenomenon based on behavioral theories.

The models developed below do not include government as a separate actor; its inclusion would complicate the formal models without significantly adding to the insights that I want to make. However, I emphasize the role of government in influencing job rights, and thus in the way the costs of adjustment are shared. I also take note of government's potential impact on the efficiency of outcomes.

The chapter concludes with a discussion of the effect of the allocation of rights on the efficiency of resource allocation. I emphasize the simple but fundamental point that multiple efficient equilibria may exist. Although an outcome may fail to maximize net monetary income of workers and firms, it may still be Pareto efficient in the sense that no one could be made better off without making someone else worse off.

A Property Rights Model of Job Security

To begin, let us consider a simple stylization of the problem facing companies in the E.C. steel industry. Firms cut back production in response to a permanent decline in demand. Either workers have strong job rights, whereby workers are guaranteed their job at their wage prior to the decline, or they have no rights to their jobs. Equivalently, either firms bear labor adjustment costs, or workers bear them.

According to the Coase Theorem, resource allocation will be independent of the assignment of property rights and, in this case, work force reductions will be independent of job rights. In neoclassical economic theory, however, there are important exceptions to the prediction that resource allocation will be independent of rights assignments. The one stressed by Coase in his seminal article (1960) results from the presence of transactions costs. If the costs of making transactions are high such that trades in job rights do not occur, then the initial allocation of rights may affect resource allocation. The definition of transactions costs can be very broadly construed to include human failure to recognize, and thus realize, beneficial trades.

In the case of the E.C. steel industry, however, such trades were common. Steelworkers accepted wage cuts or moderate wage increases in return for job security. Most companies effectively bought out workers' jobs through severance payments to those leaving voluntarily and through early retirement programs, which essentially were job buyouts of older workers. Thus, transactions costs by themselves are unlikely to account for large differences in work force adjustment due to job rights.

If there is a market for job rights, under what conditions is the amount for which workers are willing to sell their right to a job different from the amount they would be willing to pay for that right? Another exception to the Coase Theorem results from the fact that property rights assignments affect wealth. For the Coase Theorem to hold in the situation of job rights, the utility that workers derive from their jobs must be solely a function of income. Utility-maximizing workers then would seek to maximize income; a change in wealth resulting from a change in property rights assignments would not affect the income-maximizing calculus. The amount that workers would have to be paid in order to leave their jobs, therefore, is equal to the amount that workers would be willing to pay firms to keep their jobs. Because the opportunity cost of labor to the firm is the same under either condition, employment levels will be the same under the alternative definition of rights.

Yet one of the most prevalent comments from the interviews conducted for this study concerned the strong attachments workers had to their job and community. When the job not only is a means of earning income but also has elements of consumption, the assignment of property rights has an impact on employment levels in a declining industry due to wealth effects. The compensation that workers will demand to give up their jobs will be greater than the amount they would be willing to pay to keep their jobs. In effect, workers will use some of the additional wealth from their job rights to obtain greater employment stability.

This result is demonstrated formally in a simple model of labor demand by a firm experiencing a permanent decline in demand for its product. In the model, I distinguish between two types of labor adjustment costs. The first type is essentially a monetary loss. Examples include job search costs, moving or commuting costs, and a lower alternative wage, such as may result from the loss of a return on investment in firm-specific human capital.

The second type of cost relates to the job's location and work attributes that workers value.[2] This category of adjustment cost, which involves nonmonetary losses, is designed to capture a wide range of negative effects resulting from unemployment, job switch, and moving. Although much of the loss incurred in the second category may involve "nonmarketed" goods, such as friendship, it is not restricted to this type.[3]

Monetary losses are modeled simply as the difference between the wage in the declining industry, w_1, and the alternative wage less the costs of job change, w_2. Nonmonetary losses are modeled as an absolute decrease in utility, B. Working with indirect utility functions, the utility of workers who retain their jobs is

(4.1) $$V = v(w_1),$$

whereas the utility in the event of an uncompensated layoff is

(4.2) $$V = v(w_2) - B$$

$$\frac{dv}{dw} > 0; \frac{d^2v}{dw^2} < 0.$$

The firm in this model maximizes profits and is a price taker. The capital stock is assumed fixed, as are the number of hours an individual works. Thus, production is written $Q(L)$, where L denotes the number of workers. At the beginning of the period, the industry experiences a fall in demand. Firms in the industry observe this decline as a drop in product price. I assume that workers are identical in preferences and ability, and that wages are flexible. At the time of the decline, the firm is hiring $\overline{L}$ workers at $\overline{w}$ wage. If workers have rights to their jobs, the profit maximization problem for the firm is

$$(4.3) \qquad \max_{w_1,\, s_1,\, L} . \quad PQ(L) - w_1 L - s_1(\overline{L} - L),$$

subject to

$$V \geq v(\overline{w}), \text{ for all } \overline{L}$$
$$L \leq \overline{L},$$

where $v(\overline{w})$ is the initial utility of workers at $\overline{w}$, and s_1 is the severance payment the firm must make to workers it lays off.

It can easily be shown that a first-order condition for profit maximization is[4]

$$(4.4) \qquad P\left(\frac{\delta Q}{\delta L}\right) = \overline{w} - s_1.$$

If workers have no right to their jobs, the profit maximization problem becomes

$$(4.5) \qquad \max_{L} . \quad PQ(L) - (\overline{w} - s_2)L,$$

where s_2 is the pay cut workers are willing to accept to keep their jobs. With identical workers, an interior solution, and no downward wage rigidity, workers are indifferent between being employed in the declining industry and being laid off. Thus,

$$(4.6) \qquad v(\overline{w} - s_2) = v(w_2) - B.$$

The first-order condition for profit maximization is

$$(4.7) \quad P\left(\frac{\delta Q}{\delta L}\right) = \overline{w} - s_2.$$

Comparing equations (4.4) and (4.7), production and employment under alternate rights assignments differ only if $s_1 \neq s_2$. It follows immediately that $s_1 = s_2$ if only the first type of labor adjustment cost, monetary losses, exists. In this case $B = 0$, and from the constraint on utility in (4.3):

$$v(\overline{w}) = v(s_1 + w_2) \quad \text{or} \quad \overline{w} - w_2 = s_1,$$

and from (4.6):

$$v(\overline{w} - s_2) = v(w_2) \quad \text{or} \quad \overline{w} - w_2 = s_2.$$

However, with $B > 0$ and concave utility functions, $s_1 > s_2$ and employment and production will be higher when workers have rights to their jobs.[5]

Intuitively, the amount workers would be willing to pay to keep their jobs, versus the amount the firm would have to pay to get them to leave, depends on the level of job security. The assignment of rights in jobs affects workers' wealth, and drives a wedge between the compensation that workers demand for the right to their jobs and the amount they would be willing to pay for that right. The assignment of rights in jobs, therefore, affects the cost of labor in a declining industry, which, in turn, affects employment and production. Adjustment will be less the stronger the job security. With the inclusion of nonmonetary losses, resource allocation is sensitive to the assignment of property rights in jobs, and the Coase Theorem does not hold, even in a partial equilibrium framework.[6]

A Contract Model of Job Security

This section provides a simple extension of the model developed above. The purpose of the extension is twofold. First, the model in this section incorporates certain stylized facts about the E.C. steel industry and forms the basis for the estimating framework used in the empirical work on the effects of job security on employment adjustment. Second, the model brings out some interesting comparisons between the approaches used in the property rights and contract models. Although property rights and contract formulations of the decision problem are similar in neoclassical theory, they can be quite different in prospect and other behavioral theories.

In the contract model, job security in a declining industry—defined by employment levels and severance payments—is the result of efficient contracts between workers and firms. I assume that the immobility of labor and capital gives rise to quasi-economic rents, particularly in the short run, and that state-contingent contracts have not been negotiated so as to uniquely determine the division of these rents. In the context of an industry that experiences a decline in demand, how the rents are divided between labor and capital captures how the costs of adjustment are shared. Thus, the analogue to the assignment of job rights in the property rights model is the division of rents in the contract model.

In the latter model, I allow both capital and labor input to vary over the adjustment period. In the European steel industry, modernization of capital equipment, which typically involved the introduction of labor-saving technology, was an integral component of restructuring. However, technological displacement compounded the problems of displacement from depressed demand. By introducing investment in the model, one can study the tradeoffs between modernization and job security.

Production in the industry may be written as a function of labor and capital. The latter, in turn, is a function of the vintage of the capital stock and new investment:

$$Q = Q[L, K_v(\overline{K}_v, I)],$$

where Q is output, L is employment, K_v is the current level of capital and its vintage, $\overline{K}_v$ measures the capital stock and its vintage in the preceding period, and I is new capital investment.[7] The notation K_v is intended to emphasize the importance of technology embodied in the capital stock.[8]

The utility function of workers is modeled as above, where nonmonetary costs of layoff are expressed as an absolute decrease in utility.

Using indirect utility functions, the utility of a worker retaining his job is

$$V = v(w_1)$$

whereas his utility in the event of uncompensated layoff is

$$V = v(w_2) - B,$$

where w_1 is the worker's wage, and w_2 is the worker's alternative income, which may include unemployment insurance.

Workers are represented by a union. If the union has a utilitarian objective function, it maximizes total utility or, equivalently, the average utility of its membership. In this model, the firm guarantees the union some average level of utility for its members. In the model of the preceding section, job rights map into a utility level that is guaranteed to workers. Here the constraint is expressed directly in terms of the level of average utility that is guaranteed to workers. The constraint captures labor's bargaining power and the distribution of adjustment costs or income between workers and firms.

An alternative interpretation of the model is to suppose that, rather than being represented by a union with a utilitarian objective function, workers face an equal probability of being laid off. In this case, the constraint facing the firm may be taken as the expected level of utility that the firm must guarantee to individual workers. This interpretation is more similar to the concept of individual property rights developed above.

The model is also formally similar to standard models in the literature on implicit contracts. In this literature, less risk-averse firms offer workers state-contingent contracts that provide insurance against layoffs.[9] The primary motivation of implicit-contract models is to study how wages and employment levels vary over the business cycle with optimal contracts. The division of rents between workers and firms is taken as exogenous. The principal motivation of the model developed here is to study explicitly how income distribution, which is analogous to the assignment of property rights in jobs, alters resource allocation in efficient contracts.

Firms are assumed to minimize costs subject to a production constraint, an assumption reflecting the fact that E.C. companies faced production quotas that may have been binding, in addition to a constraint that they provide workers some minimum guaranteed average utility. Formally, the firm's cost minimization problem is

$$(4.8) \qquad \min_{L,\, I,\, w_1,\, s} \quad w_1 L + s(\bar{L}-L) + rI,$$

subject to:

$$Q = Q[L, K_v(\bar{K}_v, I)]$$

$$\frac{L}{\bar{L}}v(w_1) + \frac{\bar{L}-L}{\bar{L}}(v(w_2+s)-B) = \bar{V},$$

where r denotes the price of capital. Investment, I, may be expressed as an implicit function of past levels of the vintage capital stock, $\bar{K}_v$, output, Q, and labor, L, and the problem may be written

$$\min. \quad w_1 L + s(\bar{L}-L) + rI(L, K_v, Q) + \lambda\left[\frac{L}{\bar{L}}v(w_1) + \frac{\bar{L}-L}{\bar{L}}(v(w_2+s)-B) - \bar{V}\right],$$

where λ is the Lagrange multiplier on the utility constraint.

First-order conditions for cost minimization include

$$(4.9) \qquad w_1 - s + r\frac{\delta I}{\delta L} + \frac{\lambda}{\bar{L}}[v(w_1) - v(w_2+s) + B] = 0$$

$$(4.10) \qquad \frac{\delta v}{\delta w_1} = -\frac{\bar{L}}{\lambda}$$

$$(4.11) \qquad \frac{\delta v}{\delta s} = -\frac{\bar{L}}{\lambda}.$$

From (4.10) and (4.11) the marginal utility of income is equated between retained and laid-off workers, implying that incomes also are equalized: $w_1 = w_2 + s$. However, with nonmonetary losses from being laid off, those who lose their jobs are worse off. Thus, even with severance payments and optimal contracts, unemployment is involuntary.[10]

Substituting for λ in (4.9) and rearranging terms,

$$(4.12) \qquad \frac{w_1 - s - B(\delta v/\delta w)^{-1}}{r} = -\frac{\delta I}{\delta L}.$$

This condition has a familiar interpretation: the firm uses labor and capital, as determined by investment, such that the marginal rate of substitution equals the inverse ratio of their costs. The numerator of the left-hand expression is the marginal cost of workers. It equals the wage less the severance payment (which equals the worker's alternative wage, w_2), less the monetary value of the disutility a worker experiences in being laid off. The marginal cost of new capital investment, which is in the denominator, is r.

The effect of labor's bargaining power, and implicitly of job rights, on resource allocation can be seen by examining the effects of changes in $\overline{V}$ on optimal adjustment. The appendix to this chapter shows these comparative static results. An increase in $\overline{V}$ will result in higher levels of employment, higher compensation, and less investment and capital. Workers will gain higher utility in part through greater job security and in part through higher wages and severance payments. An increase in B, the nonmonetary costs of adjustment, also will result in more employment and less capital investment; however, the effect on wages is ambiguous. With an increase in nonmonetary costs of adjustment, workers may choose to take a wage cut in order to gain greater job security. The allocation of labor and capital in a declining industry, therefore, depends on the nature and magnitude of the adjustment costs to labor and on the division of the adjustment costs between labor and capital.[11]

The Role of Government

In the European steel industry, governments played an important role in devising and financing work force reduction programs, both in publicly owned companies and as third parties in the negotiation process. A direct effect of government assistance was to help define or protect workers' job rights and job security. Many companies could not have afforded to give strong job guarantees to their workers without outside help. The government may act as an insurer of job rights against a firm's bankruptcy—a role that has counterparts in banking and other financial industries. Here, the role of government may be seen as influencing job rights and the distribution of the costs of adjustment, which are taken as exogenous in the above models. Government subsidies of various measures shifted much of the burden of adjustment onto the public sector.

In addition to having distributional impacts, government assistance may have had important effects on the efficiency of the adjustment process. If layoffs from a declining industry have negative spillover effects on a local economy, then government subsidies of job security measures may help internalize the externality.[12] Of course, ill-conceived subsidies may distort labor prices and cause insufficient adjustment of work force levels. These sorts of efficiency effects are not captured in the above models, which assume that adjustment is efficient for any given distribution of rights or income.

Job Rights and Preferences

In the property rights and contract models, the assumption that workers place a value on their jobs independent of the income these jobs provide underlies the result that job rights, or the level of utility guaranteed to workers, affect employment levels. In the absence of transactions costs, such income effects from rights assignments are necessary to yield these results in neoclassical economic theory. However, recent empirical evidence suggests that even controlling for income effects, the initial assignment of property rights influences decisions. The compensation demanded for that right is significantly greater than the willingness to pay, contrary to predictions of neoclassical economics.

Some of the most direct evidence comes from experimental studies reported in Knetsch and Sinden (1984). In a series of experiments, participants either were offered the opportunity to purchase a lottery ticket at a set price or were given a lottery ticket and offered the chance to sell it at that same price. Income effects from these rights to lottery tickets were minimal because the price of the ticket was relatively low and, in certain experiments, because participants who did not receive lottery tickets were given cash equivalents. In almost all of the experiments, the number of individuals willing to buy the ticket was significantly lower than the number of individuals willing to sell, implying a much lower average value under the willingness-to-pay measure than under the compensation-demanded measure.

Various explanations have been offered to account for this phenomenon. Knetsch and Sinden suggest that it is related to cognitive biases against a feeling of regret resulting from a deliberate change in asset position or against the greater mental effort required in evaluating the costs and benefits of a change.

The phenomenon also may be explained by the tendency of individuals to treat losses and gains quite differently. According to prospect theory, developed by Kahneman and Tversky,[13] a loss has greater impact on utility than does an equivalent gain. Consequently, decisions are sensitive to an individual's reference point. The framing of a problem, in turn, may affect that reference point, and thus may affect whether an individual treats a certain choice as involving a loss or a gain. According to Knetsch and Sinden, the person who has a property right in the lottery ticket treats the money from the sale of the ticket as a gain, whereas the person who does not treats the payment for the ticket as a loss.[14]

Evidence from these experiments is consistent with a more general phenomenon in which the framing of the decision problem affects the decision. The asymmetric treatment of losses and gains coupled with framing effects has been used to understand a number of phenomena unexplained by neoclassical economics. For example, if a price differential exists (if, say, gasoline costs more when purchased by credit card than when purchased with cash), consumers are more likely to pay the higher price when the differential is labeled a discount rather than a surcharge. One of the most commonly cited economic phenomena that can be explained by this theory is "money illusion," in which workers find the lowering of real wages through inflation more acceptable than nominal wage cuts.[15]

In the case of job rights, this theory predicts that workers would demand greater compensation for leaving their jobs than they would be willing to pay to keep their jobs. To illustrate, consider the simple definition of job rights used earlier in this chapter and a hypothetical experiment similar to that reported in Knetsch and Sinden. Suppose that in one group, workers have rights to their jobs and may be bought out for an amount s. In another group, workers have no rights to their jobs; however, their income is higher than the first group's by the amount s, and they may accept a wage cut of this amount to keep their jobs. Where workers have rights to their jobs, s is framed as a gain or an opportunity cost. Where workers do not have rights to their jobs, s is treated as a loss. Neoclassical theory would predict the same employment levels in either situation. Results consistent with prospect theory and with the experimental evidence of Knetsch and Sinden would show employment levels higher where workers have an explicit property right in their jobs.

Property rights may affect in a fundamental way how workers value jobs. In essence, the disutility workers experience when laid off may be less if they expected they would be laid off in a situation of declining demand than if they expected they would be retained in that situation. The disutility experienced by workers who are laid off—and the industrial relations problems a firm experiences when it lays off workers—is likely to depend on the expectations of those workers arising from explicit or implicit job rights.

Job Security, Efficiency, and Income Distribution

The theories discussed above offer different and to some extent complementary explanations of why job rights might affect the allocation of resources in a declining industry. Common to all of the theories is the fact that worker rights to job security may affect work force reduction and investment decisions, and may, in a popular sense, impede restructuring; but this outcome, in itself, does not imply an allocation of resources that is Pareto inefficient.

When workers have strong attachments to workplace and community, the assignment of rights in jobs or the division of rents affects employment levels in efficient contracts. According to neoclassical economic theory, this is because with a higher endowment of job rights or rents, workers will choose more employment stability. In a declining industry, the reduction of work force levels as well as investment in capital is less the stronger workers' job rights or the greater labor's share of the economic rents.

Implicit in the models presented in this chapter is the assumption that there is no unique equilibrium arising from competitive markets that defines wages and levels of job security. Labor and capital are imperfectly mobile, particularly after a worker's job has been selected or capital's use determined. The immobility of labor and capital creates a bilateral monopoly situation between workers and firms in which the division of rents, or the costs of adjustment, are not uniquely determined.[16]

Prospect and other behavioral theories offer quite different explanations for why rights assignments may affect economic outcomes. At the heart of these theories is the proposition that the assignment of rights affects underlying preferences. Workers may appear to value a job more the stronger their right in that job, because that right alters their reference point. As a result, firms may have to compensate workers more to give up their jobs than workers would be willing to pay to keep their jobs, not only because rights assignments effectively change workers' wealth, but also because the assignment of rights affects workers' underlying preferences.

Such instability of preferences has interesting implications for the relationship between the property rights and contract models. Under standard assumptions of neoclassical economic theory, the assignment of job rights between workers and firms is analogous to the division of rents between labor and capital, and the theoretical formulations can be shown to be equivalent.[17] However, if preferences are unstable, the property rights and contract models would not yield the same results, because property rights cannot be translated into an equivalent level of utility without affecting outcomes. The equivalence of the outcomes under the two formulations depends on a stable set of preferences.

All of the theories discussed in this chapter point to the existence of multiple, Pareto-efficient equilibria, which vary with the assignment of property rights or the division of rents between labor and capital. The emphasis on distributional issues is not intended to deny the existence of market inefficiencies in particular instances of adjustment in the E.C. steel industry. Government programs that subsidize employment or capital may have affected private choices in inefficient ways. Private firms or government-owned companies may have failed to fully exploit efficient mechanisms for reducing work force levels.

This discussion, however, does seek to draw attention to other important issues. Job security is inextricably linked with the distribution of income and risk in society and with the values that society places on employment and community stability. The theories presented in this chapter have been designed to elucidate these choices and their consequences. Strong worker rights in jobs and thus job security may result in substantially fewer work force reductions, which, in turn, may affect the amount and pattern of investment and disinvestment in a declining industry.

Appendix

Using the fact that $w_1 = s + w_2$, the first-order conditions for cost minimization from (4.8) may be written:

$$(4A.1) \quad w_2 + r\frac{\delta I}{\delta L} + \frac{\lambda B}{\overline{L}} = 0$$

$$(4A.2) \quad 1 + \left(\frac{\lambda}{\overline{L}}\right)\left(\frac{\delta v}{\delta w}\right) = 0$$

$$(4A.3) \quad v(w_1) - \left(\frac{\overline{L} - L}{\overline{L}}\right)B - \overline{V} = 0.$$

The second-order condition for cost minimization is

$$(4A.4) \quad -\left(\frac{r}{\overline{L}}\right)\left(\frac{\delta v}{\delta w}\right)^2\left(\frac{\delta^2 I}{\delta L^2}\right) - \left(\frac{\lambda}{\overline{L}^3}\right)\left(\frac{\delta^2 v}{\delta w^2}\right)B^2 < 0.$$

Noting that $\lambda < 0$, this second-order condition is satisfied.

The comparative static results for an increase in $\overline{V}$, the average utility guaranteed to workers, are

$$(4A.5) \quad \frac{\delta L}{\delta \overline{V}} = \frac{-\left(\frac{\lambda B}{\overline{L}^2}\right)\left(\frac{\delta^2 v}{\delta w^2}\right)}{D} > 0$$

$$(4A.6) \quad \frac{\delta w}{\delta \overline{V}} = \frac{-\left(\frac{\delta v}{\delta w}\right)\left(\frac{\delta^2 I}{\delta L^2}\right)r}{D} > 0,$$

where D is the expression in (4A.4), which is negative. An increase in $\overline{V}$ will result in more labor, and by implication less investment and use of capital, and in higher wages.

The comparative static results for an increase in B, the nonmonetary costs of layoffs, are

$$(4A.7) \quad \frac{\delta L}{\delta B} = \frac{\left(\frac{\lambda}{\overline{L}^2}\right)\left(\frac{\delta v}{\delta w}\right)^2 - \left(\frac{\lambda(\overline{L} - L)B}{\overline{L}^3}\right)\left(\frac{\delta^2 v}{\delta w^2}\right)}{D} > 0$$

$$\text{(4A.8)} \quad \frac{\delta w}{\delta B} = \frac{-\left(\frac{\bar{L}-L}{\bar{L}^2}\right)\left(\frac{\delta v}{\delta w}\right)\left(\frac{\delta^2 I}{\delta L^2}\right)r - \left(\frac{\lambda B}{\bar{L}^3}\right)\left(\frac{\delta v}{\delta w}\right)}{D} \lesseqgtr 0.$$

Both expressions in the numerator of (4A.7) are negative, implying that the entire expression is positive. Thus, an increase in nonmonetary costs of adjustment will result in fewer layoffs and less investment in new capital. However, the first expression in the numerator of (4A.8) is negative, while the second expression is positive. Thus, the effect of an increase in B on wages is ambiguous. With an increase in nonmonetary adjustment costs, workers may choose to take wage cuts in order to gain even more job security. Whether wages increase or decrease will depend upon the relative magnitude of B and the substitutability of labor and capital.

5

Job Security and the Adjustment of Employment in Steel

The most direct effects of job security in a declining industry are on the adjustment of labor. According to the theoretical discussion in the preceding chapter, stronger job security should result in less reduction of work force levels for any given decline in production. This chapter examines the empirical evidence of such effects on the adjustment of labor in the European Community steel industry. Job security potentially has more pervasive influence on restructuring in a declining industry, affecting production, investment, and closure decisions. The following chapter deals with these broader effects.

As a result of differences in industrial relations institutions and government policy, substantial differences in the level of job security existed across the countries covered in this study. The empirical approach adopted here involves testing for cross-country differences in employment adjustment that are consistent with observed levels of job security, using detailed plant data. Plant-level data are supplemented with aggregate country data to assess the extent to which cross-country differences in the adjustment of employment levels are accounted for by differences in the adjustment of average hours per worker.

The job security afforded workers arguably is, in part, a function of the cost pressures to restructure: job security will tend to be weaker and work force reductions greater where initial productivity was lower. Consequently, observed differences across countries in the adjustment of employment may be due, in a causal sense, not to differences in worker rights to job security but rather to underlying factors, such as average differences in productivity. Indeed, Britain, the country with the weakest job security, also had relatively low productivity, owing to poor labor relations and antiquated plant and equipment. Tests of cross-country differences in employment adjustment endeavor to sort out the direction of causality by controlling for productivity, the vintage of the capital stock, and other factors that jointly may determine job security and work force reductions.

The Estimating Framework

Firms in the steel industry are assumed to minimize production costs subject to the constraint that they guarantee workers some average level of utility. The model underlying the empirical estimation was developed in the preceding chapter. Equation (4.8) is repeated here.

$$\min_{L,w_1,s,I} \quad w_1L+s(\overline{L}-L)+rI \tag{5.1}$$

subject to

$$Q = Q(L,K_v(I,\overline{K}_v))$$

$$\frac{L}{\overline{L}}v(w_1)+\frac{\overline{L}-L}{\overline{L}}(v(w_2+s)-B) = \overline{V}.$$

Capital is a function of new investment and of the initial capital stock and its vintage.

Employment, one of the choice variables for the firm and the focus of the empirical investigation in this chapter, may be written as an implicit function of the exogenous variables in equation (5.1).

$$L = f(\overline{L},Q,\overline{K}_v,w_2,B,r,\overline{V}).$$

In the model, past levels of employment, capital, and output are taken as exogenous. Therefore, $\overline{L}$ may be expressed as a function of past output and the capital stock, and employment adjustment may be written

$$L-\overline{L} = g(Q,\overline{Q},\overline{K}_v,w_2,B,r,\overline{V}), \tag{5.2}$$

where $\overline{Q}$ is production in the preceding period.

A firm's optimal adjustment of employment to a decline in production depends upon the size of the decline, $Q - \overline{Q}$; the quality and vintage of its capital stock, $\overline{K}_v$; the external labor market conditions, captured by w_2; the nonmonetary costs of layoff, B; the price of new capital, r; and the employment security obligations of the firm to its workers, $\overline{V}$.

The technology embodied in the capital stock may affect the degree of variability of labor in the production process, which, in turn, would affect the elasticity of labor with respect to production, even in the absence of new investment. The quality and vintage of the capital stock also may affect employment adjustment during restructuring by affecting the level of new investment. With everything else constant, one would expect that the older the capital stock, the higher the marginal productivity of new investment. From this fact, one would expect older plant and equipment to be associated with more investment and consequently greater reductions in employment. In steel, older capital stock tends to be more labor intensive. Therefore, an older capital stock usually implies greater worker displacement from any new investment, a factor that may mitigate capital labor substitution, particularly during a downturn.

Higher regional unemployment and other labor market factors that would adversely affect a worker's alternative wage would be expected to result in fewer work force reductions. For the purposes of estimation, the nonmonetary costs of layoff and the cost of new capital investment are assumed to be the same across workers and firms within Community countries.

In the model, $\overline{V}$, which captures worker rights to job security, is taken as exogenous. Its determination is a complex process that depends on industrial relations factors, legal institutions, and government policy. For example, job security laws strengthen the position of labor in negotiating terms of layoff. Special industrial relations institutions, such as that of codetermination in the German steel industry, may place labor in a strong position to block layoffs. Government programs, including special assistance to the steel industry, may affect how a firm optimally adjusts its employment levels.

The legal, political, and institutional variables underlying $\overline{V}$ in equation (5.1) were similar within countries, but varied substantially across countries. Such differences may have resulted in the quite different levels of job security and patterns of employment adjustment that one observes, particularly between the Continental countries and Britain.

The approach here involves testing whether there were different patterns in the adjustment of employment levels across countries during the steel crisis, after controlling for other relevant factors in (5.1). The null hypothesis is that the level of job security of steelworkers was simply a function of the other variables in equation (5.1). What we observe empirically are not exogenous job rights, but rather the outcomes of a complex negotiating process. The weaker job security of steelworkers in Britain, for example, could have resulted from less efficient capital stock and the need to modernize, rather than from any inherently weaker job rights possessed by labor.

The Data

To estimate employment adjustment equations, I collected plant-level data on production, employment, and the capital stock, as well as regional data on labor market conditions. The data base also includes wages and product prices. Below is a description of each data item and of the problems of matching data from different sources. Background on the production process in the steel industry is provided where this is necessary to understand the data. Table 5.1 lists each data item, its level of aggregation, and the period of coverage.

Plant-Level Data

The core of the data base for this study consists of plant-level data on production, capacity, employment, and various measures of the quality and vintage of the capital stock. Although data on production, capacity, and employment are available for all works in the European Community, only those works for which corresponding data on the capital stock were available are used in the empirical analysis. Specifically, the data base covers eighty-eight works; it incorporates all of the major integrated producers and accounts for 80 to 85 percent of production and capacity in the European Community.

The production of steel products comprises a number of stages. In an integrated plant, the first stage involves the production of iron. Coke ovens convert metallurgical coal into coke. In some cases, additional processing is required to produce a higher grade of iron ore. Coke and iron ore, along with limestone, then are combined in the blast furnace to produce pig iron. In the next stage, steel making, the molten iron is transformed into steel ingots, and steel ingots then are made into semifinished products (slabs, blooms, or billets). In the continuous-casting process, liquid steel is cast directly as a semifinished product. Semifinished products are then rolled or otherwise formed into a wide variety of products in hot-rolling mills. Hot-rolled products may receive further processing, such as cold-rolling, pickling, and galvanizing, or they may be cut as in the production of plate and of hoop and strip from hot-wide strip.

The data base includes detailed information on the various stages of the production process. Annual production data are available from

Table 5.1. Summary of data base.

Data item	Aggregation level	Period of coverage
Production and capacity	*Works*	1974–1982, 1986[a]
Iron		
Steel, total		
Open hearth		
Continuously cast		
Electric		
Hot rolling, total		
Plate		
Hot-wide strip		
Heavy and medium sections		
Hoop and strip		
Merchant bars, concrete reinforcing bars, and wire rod		
Finished products		
Capital stock	*Installation*	1977, 1982
Blast furnace		
Diameter		
Year of construction		
Year of last modernization		
Capacity		
Index of technical efficiency[b]		
Steel making		
Year of construction		
Year of last modernization		
Capacity		
Index of technical efficiency[b]		
Hot rolling		
Year of construction		
Year of last modernization		
Capacity		
Index of technical efficiency[b]		
Employment	*Works*	1974, 1977–1982
Regional labor market conditions	*Region III*	1979
Unemployment rates		
Dependency on steel		
Wages	*Region II*	1975–1981
Prices	*Average country prices for individual hot-rolled products*	1974–1982

a. 1986 data available only for capacity at the hot-rolling stage.
b. Index available only for 1982.

1974 to 1982 at each stage of production: iron making, steel making, hot-rolling, and finished goods. The steel-making stage is broken down into continuously cast steel, steel produced in open-hearth furnaces, and steel produced in electric furnaces. At the hot-rolling stage, production is broken down by product group: hot-wide strip, plate, sections, hoop and strip, wire rod, merchant bars, and concrete reinforcing bars. Since production of the last three categories may occur in the same mill, they were aggregated in order to match them with capital stock variables.

Data on capacity are available at the same level of aggregation as the production data from 1974 to 1983. In addition, the data base includes the negotiated capacity levels by detailed product group at the hot-rolling stage for 1986.[1] An accurate measure of capacity is difficult to construct, since much equipment, though still intact, may be inactive, and may require considerable upgrading and time to be brought on line. The definition of capacity used by the E.C. is "maximum possible production," which counts only capacity that is active or that could be quickly activated. All data on production and capacity are expressed in physical units.

Annual employment data are available at the works level for 1974 and from 1977 to 1982. Employment data are missing for Britain for the years 1979 and 1981, and for Italy, the Netherlands, and Luxembourg for the year 1981. Employment is measured at the end of the year.

A number of data items were collected to capture important elements of the quality and vintage of the capital stock. Since World War II, innovations in the production process have radically altered the optimal equipment, scale of operation, and layout in the steel industry. In the hot-rolling stage, important technical advances have been made in the speed of operation, the degree of automation, and the quality of the surface. Some of the most important innovations have occurred in the steel-making stage. The basic oxygen furnace (BOF), first introduced in 1950, requires much shorter heat times than the open hearth or the Bessemer process, and thus drastically reduces the capital, labor, and energy input per unit of output. Continuous casting, which involves directly casting liquid steel into semifinished shapes, increases productivity, particularly through energy savings.

Accompanying the replacement of the open hearth by the BOF has been an increase in the use of the electric furnace. Electric furnaces usually rely entirely on scrap, rather than on molten pig iron, for their charge, and thus their cost-effectiveness depends largely on the price of scrap. The growth in scrap supply in the postwar period, along with the introduction of the BOF (which, unlike the open hearth, uses virtually no scrap), lowered scrap prices and spurred widespread adoption of electric furnaces.

Technical economies of scale exist in two areas: the blast furnace and the hot-wide strip mill. The minimum efficient scale for a blast furnace is a diameter of between ten and twelve meters, with a production of between 2 and 3 million tons per year. The hot-wide strip mill has a minimum efficient scale of between 3 and 4 million tons (Crandall 1981, and Montenero 1982). Taking into account economies of scale and indivisibilities in other equipment, estimates of the minimum efficient scale for an integrated steel works vary from 4 to 8 million tons, according to whether the plant's product range is narrow or wide (Adams 1982, p. 87).[2]

Although there are substantial scale economies for the integrated producer, this is not the case for the so-called mini-mill. Mini-mills use electric furnaces, and thus bypass the large capital expense and economies of scale in iron production. Furthermore, they do not produce hot-wide strip, the basic product of the steel industry used to produce automobiles, appliances, and other consumer and capital goods. Rather, they tend to concentrate production in certain product lines—in particular, wire rod and small merchant bars. The empirical work in this study excludes mini-mills.

The data on capital stock used in this study come from two sources. The first is a data base underlying a simulation model used by the E.C. Commission to evaluate the financial viability of a works. It contains detailed information on equipment in works in the sample as of 1982. Items extracted from this data base include the year of construction, year of last modernization, capacity, and classification as active or inactive for each installation in the works, from the blast furnace through the hot-rolling mills. It is assumed that the year of construction and year of last modernization would largely capture obsolescence resulting from process innovations. In addition, data on the diameter of the blast furnace were collected. Finally, each installation has a rating of its technical efficiency; the scale goes from 1 to 5, with 5 being the most efficient. The ratings were developed for the E.C. Commission from algorithms based on a much larger set of technical variables, which varied according to the installation.

It was desirable to have these measures of the capital stock for some point in time representing the beginning of the restructuring; that is, the time when firms began making investment, disinvestment, and employment reduction decisions based on new information of the structural nature of the decline in demand and excess capacity in the European industry. Because of potentially large changes in the capital stock during a period of restructuring, the 1982 data were back-dated to take into account additions, closures, and modernization. The data source was a EUROSTAT survey of plant equipment as of the end of 1977.

Although the choice of the base year was governed, in part, by data availability, measuring the capital stock with some lag from the time of the downturn is justifiable. The industry was slow to recognize the structural nature of the crisis. The E.C. Commission did not declare the crisis to be structural until the end of 1977, and widespread acceptance of its structural component in the industry came somewhat later. Furthermore, because of long execution lags in the steel industry, most changes in the capital stock occurring between the end of 1974 and the end of 1977 would have been based on decisions made during the boom period. The same data items, with the exception of the equipment classification variables, were collected for any installation that either was missing in the 1982 data base or that had been modernized after 1977. The 1977 survey also furnished a consistency check for the 1982 data.

The selection of 1977 as the base year also means that any works closed before the end of 1977 was excluded from my sample. This is particularly important in the case of Britain, where the British Steel Corporation closed a number of plants between 1974 and 1977. Although certain closings were accelerated by the crisis and were controversial, these plants had been targeted for closure by a restructuring plan developed prior to the downturn, and I thought it inappropriate to include them.

In constructing the data base, it was also necessary to develop a summary statistic for certain capital stock variables. Often a works has more than one installation for producing a product type, particularly at the iron- and steel-making stages. When this was the case, the median value of all capacity classified as active was used to match it with production data. For example, all fully integrated works have more than one blast furnace. Therefore, in the data base a value of, say, nine meters for the item "diameter of blast furnace" implies that 50 percent of all blast furnace capacity classified as active in 1977 has a diameter of nine meters or more.[3]

Data on Regional Labor Market Conditions

In addition to data at the plant level, I collected regional data on labor market conditions that would be expected to affect the opportunity cost of labor, and thereby affect employment adjustment and production and capacity allocation decisions. The European Community regional classification system comprises three levels. Level I represents the country. Level III, the least aggregated, corresponds roughly to the U.S. county classification. The location of each works in the sample was determined at the level III region.

Variables that capture labor market conditions are regional unemployment rates and measures of the region's dependence on steel for employment. Because of differences in definitions of unemployment and in social security systems, unemployment rates across countries are not comparable. The E.C. Commission has constructed unemployment rates at the level III region for the years 1979 and 1981 that attempt to adjust for differences in defintions.[4] Although problems of comparability remain, these unemployment figures are used.

The European Commission sponsored a study to determine a region's dependence on steel for employment. The study defines a level III region as "highly dependent" on steel if steel employment accounted for at least 10 percent of industrial employment in 1979. Other sites were added to this list, taking into account a region's size and local concentrations of steel. This classification is used as a dummy variable in the empirical analysis.

Wage and Price Data

Annual industry wage data at the regional level are contained in a EUROSTAT publication titled *Regional Statistics*. The relevant industry classification for steel also includes the production and first transformation of nonferrous metals.[5] List prices for individual hot-rolled products in each country were used to develop an average product price for each plant according to its hot-rolled product mix.[6] This product price, in turn, was used to deflate the wage data. Because of price wars and numerous allegations of cheating on list prices during the period, these prices generally are regarded as unreliable. Despite the unreliability of absolute price levels, it is assumed that the data accurately reflect price trends in the countries. However, because of these problems, regional wages also were deflated by consumer price indices as an alternative wage measure.

The Adjustment of Work Force Levels

This section examines the adjustment of employment levels over the 1974–1982 period and the 1974–1977 subperiod. During this time, the steel industry was characterized by large declines in production and extensive modernization. The empirical analysis focuses on identifying systematic differences in the adjustment of employment levels across countries that are consistent with the job security policies outlined in Chapter 3.

Job security policies within countries were similar across plants in the sample. In two of the countries, Britain and Luxembourg, only one company is represented. In France and Italy, social plans governing job security negotiated at the national level covered all of the plants in the sample. In Germany and Belgium, where policies were formed at the company or plant level, similar levels of job security may be attributed to a common social environment and government programs that subsidize short-time work and early retirement. In Germany, although compensation levels for items like early retirement did differ significantly across plants, basic job guarantees did not.

Cross-country differences in job security and employment adjustment may result from differences in social, political, and legal factors that affect worker rights in jobs or labor's bargaining power. Alternatively, they may be due to differences in economic conditions across steel regions or to differences in the competitive position of the industry across countries during the crisis.

To test the relative merits of these hypotheses, the relationship in equation (5.2) is estimated using the following functional form:

$$\begin{aligned}\Delta\log L_i = {} & \beta_0 + \beta_1 \Delta\log Q_i + \beta_2 \text{GROW} * \Delta\log Q_i \\ & + \beta_3 w_{2i} + \beta_4 \overline{K}_{vi} + \sum_{\text{j}} \beta_{5j} C_j + \varepsilon_i .\end{aligned}$$

The dependent variable is the change in the log of employment. Independent variables include the change in the log of production; the interaction of the change in the log of production with a dummy variable, GROW, capturing whether production in the plant expanded or contracted during the period; regional labor market conditions that would affect steelworkers' employment opportunities, w_2; measures of the quality and vintage of the capital stock, $\overline{K}_v$; and country dummy variables, C_j. The subscript i indexes the plant; ε is the error term.

Two variables are used to measure labor market conditions and the opportunity cost of steelworkers: a dummy variable that equals 1 if a commission study classified the region in which the works is located as highly dependent on steel for employment; and the regional unemployment rate for 1979.

Measures of the quality and vintage of the capital stock include the median year of construction of the works' hot-rolling equipment, the median year of last modernization of its hot-rolling equipment, the median year of construction of its steel-making equipment, and the median diameter of its blast furnace capacity.

In addition, 1974 employment in the works was divided by its production in that year as a crude measure of initial productivity. The figure is not adjusted for capacity utilization, though the industry was operating near full capacity that year. Differences in this productivity measure capture, in part, the age of the capital stock and the labor intensity of the production process. This measure, therefore, supplements the capital stock variables.

The variable also may capture technical inefficiency due to the inefficient utilization of labor for any given capital stock. Though not a part of the formal theory developed above, it is commonly believed that numerous steel works were grossly overstaffed prior to the crisis. According to the theory of X-efficiency developed by Leibenstein (1966), individuals may not behave as utility maximizers, and thus firms may not behave as profit maximizers. Under competitive pressure, however, firms reduce their organizational slack.[7] The increase in competitive pressure in the steel crisis may have compelled inefficient firms to increase labor productivity. The British Steel Corporation, for instance, pointed to such inefficiencies to argue the inevitability of quick, massive work force reductions.

The model of employment adjustment was developed in the context of a firm experiencing a decline in production, and the adjustment costs are specific to that situation. Although most of the plants in the sample experienced a decline in production over the period, for a significant minority production actually increased. To allow for different adjustment costs and thus employment elasticities for these plants, the regressions include the interaction of the change in the log of production with a dummy variable that equals 1 if the plant's production increased over the period.

Country dummy variables are included to test for cross-country differences in the adjustment of employment levels that result from underlying differences in job rights of workers. In countries with relatively strong job security, one would expect less reduction of employment levels, and thus a positive coefficient on the country dummy variable. A positive coefficient on a country dummy variable would capture the adjustment of average hours per worker in lieu of employment levels or a decline (or slower growth) in labor productivity over the period.[8] If differences in job security and work force adjustment across countries are due to differences in regional labor market conditions and the competitive position of the industry, then the coefficients on the country variables should be insignificantly different from zero. A rough dichotomy in the levels of job security existed between Britain, where workers had weak job security, and the Continental countries, where workers received strong protection against layoff. For ease of comparison, then, Britain is the omitted country. The Netherlands is excluded from estimation, since there is only one steel works in that country.

To save degrees of freedom, the output term is defined as the sum of the production of iron, crude steel, and hot-rolled products.[9] In the model developed above, output is assumed to be exogenous. Yet this may not be the case. By affecting the costs of work force adjustment, the level of job security also may affect production decisions; works with strong job security and relatively fewer employment reductions may have less reduction in output. In light of a possible simultaneity bias, both ordinary least squares (OLS) and nonlinear two-stage least squares (2SLS) models were estimated.[10] Although Hausman tests suggest that simultaneity bias is not a problem for the OLS estimates, both sets of coefficients are reported in the tables below. Chapter 6 deals explicitly with determinants of the allocation of production across plants during this period.

The analysis includes only plants that remained open. There were relatively few closures during this period: four in Britain and two in France. Results on employment adjustment, however, should be interpreted as conditional on the works remaining open.[11]

Because steel policies that affected restructuring plans and steelworkers' level of job security changed over time in certain countries, analysis of employment adjustment over various time horizons is of interest. Unfortunately, data problems limit such comparisons. Employment data are missing for Britain for 1979, and for Britain, Italy, Luxembourg, and the Netherlands for 1981. In addition, strikes in Britain in 1980 and in Germany in 1978–1979 distort the employment picture for these years. Finally, no data on employment are available for the years 1975 and 1976.

Tables 5.2 and 5.3 and Tables 5.4 and 5.5 present evidence on cross-country differences in employment adjustment over the 1974–1979 and 1974–1982 periods. The equations in Tables 5.2 and 5.3 include controls for regional labor market conditions, and those in Tables 5.4 and 5.5 add controls for the quality and vintage of the capital stock.

In 1977 there still was considerable uncertainty over the structural nature of the decline and its severity. Important decisions to restructure had not yet been made, for the most part. In Britain, the Labour party was in power and was using the nationalized steel industry to bolster employment. In addition, British steel initially did not experience so severe a decline in production as did the Continental countries. Although substantial restructuring did occur in Britain in these early years, the BSC was concentrating its efforts on plants targeted for closure prior to the downturn. These plants are excluded from the sample. In France, major restructuring and consolidation under the Barre Plan began in 1978. Only in Luxembourg did management, labor, and government officials make an early decision to move workers out of steel, primarily to special government projects.

Reflecting this fact, the coefficient on the Luxembourg country dummy variable in Table 5.2 is large, negative, and statistically significant, indicating that employment adjustment was significantly greater in Luxembourg than in Britain. Tests also show that employment adjustment generally was significantly greater in Luxembourg than in the other Continental countries. The coefficients on the other country dummy variables are generally positive, indicating that during this period there was less reduction of employment in these countries than in Britain for a given change in output. However, they are generally insignificant, except in the case of Italy.

In 1980 the Conservative government in Britain instituted a new restructuring plan for its steel industry and significantly weakened the job security of steelworkers. The results for the 1974–1982 period, correspondingly, show quite strong differences in the adjustment of employment levels between Britain and the Continental countries. In Table 5.3 the coefficients on all of the country dummy variables are positive, indicating that, for any given change in output, work force reductions were less in these countries than in Britain.[12] The coefficients on the dummy variables for Germany, France, Belgium, and Italy are large and statistically significant in all equations. For example, the equations imply that, controlling for production and regional labor market conditions, German steel plants reduced work force levels on the order of 50 percent less than did British steel plants.

The coefficients on the Luxembourg term, though positive, are lower than the coefficients for Germany, Belgium, and Italy.[13] The relatively large employment adjustment in Luxembourg implied by these results over the 1974–1977 and 1974–1982 periods is consistent with its strong job security, and may be explained by its Anticrisis Division. Many workers were subcontracted out and were no longer counted in the steel sector.

In addition, though the coefficients on the French dummy variable in the equations reported in Table 5.3 are positive and statistically significant, they also are generally significantly lower than in Germany, Belgium, and Italy. Given the strong job security in France during the period, this result, which indicates greater adjustment of employment over the period in France than in other Continental countries, is somewhat surprising. Several factors account for its relatively large adjustment. First, the regions in France that underwent the greatest restructuring, Lorraine and the Nord, were also the regions with by far the largest concentrations of foreign steelworkers. The French government had a program in which it paid foreigners to leave the country. Examination of data from 1977 to 1979, the years for which I have regional breakdowns of foreign workers in France, shows that in the Nord region, where workers from non-E.C. countries accounted for about 14 percent of the work force in 1977, they accounted for over 20 percent of the net work force reductions over the period; in Lorraine, where they accounted for 27 percent of the work force, they absorbed more than 50 percent of the net reductions. In addition, in 1978 the industry operated an extremely popular job buyout program, open to all workers.

Probably the most important factor accounting for the large work force reductions in France relative to other Continental countries was its extensive use of early retirement. By 1979 workers were going on early retirement at age fifty. According to union figures, from January 1979 to November 1982 more than 32,000 French steelworkers left on early retirement, and more than half of these were younger than fifty-five.[14] Thus, the rapid employment reductions in France mirror the industry's aggressive use of early retirement and other forms of job buyout.

Apart from cross-country differences, one of the most striking

Table 5.2. Employment adjustment, 1974–1977.[a]

Variable	Mean[b]	(1)[c]		(2)[c]		(3)[c]	
		OLS	2SLS	OLS	2SLS	OLS	2SLS
Constant	n.a.	.016	.059	.018	.018	.016	.073
		(.036)	(.058)	(.091)	(.099)	(.050)	(.078)
Change in ln(output)	−.314	.462[d]	.543[d]	.460[d]	.591[d]	.459[d]	.565[d]
	(.376)	(.052)	(.106)	(.057)	(.157)	(.055)	(.156)
*Growth	.043	−.516[d]	−.773[d]	−.517[d]	−.871[d]	−.516[d]	−.827[d]
	(.173)	(.113)	(.224)	(.118)	(.313)	(.115)	(.314)
West Germany	.315	.062[e]	.050	.063	.071	.062	.047
	(.468)	(.037)	(.039)	(.050)	(.056)	(.037)	(.042)
France	.247	.067[e]	.067	.067	.060	.066	.066
	(.434)	(.039)	(.041)	(.040)	(.045)	(.040)	(.043)
Belgium	.055	.013	.001	.012	−.019	.010	−.003
	(.229)	(.064)	(.067)	(.069)	(.078)	(.068)	(.072)
Luxembourg	.068	−.176[d]	−.189[d]	−.174[d]	−.151[e]	−.177[d]	−.192[d]
	(.254)	(.059)	(.062)	(.081)	(.091)	(.062)	(.067)

Italy	.068 (.254)	.121[d] (.060)	.102 (.065)	.119[e] (.063)	.100 (.069)	.118[e] (.062)	.095 (.069)
Regional dependence on steel for employment	.781 (.417)	—	—	−.006 (.038)	.000 (.045)	−.006 (.038)	−.005 (.044)
Regional unemployment	6.16 (2.27)	—	—	.0002 (.011)	.009 (.015)	—	—
Change in product wage	.678 (.428)	—	—	—	—	.005 (.035)	.001 (.039)
R^2		.650	.623	.651	.600	.651	.610

a. Dependent variable: change in ln(employment). Its mean (standard deviation) is −.118 (.186). A dash appearing in a column indicates that the variable was not included in the regression equation.

b. Standard deviations are in parentheses.

c. Standard errors are in parentheses. Number of observations is 73. OLS = ordinary least squares model. 2SLS = two-stage least squares model.

d. Coefficient significant at the .05 level or better.

e. Coefficient significant at the .10 level.

Table 5.3. Employment adjustment, 1974–1982.[a]

Variable	Mean[b]	(1)[c] OLS	(1)[c] 2SLS	(2)[c] OLS	(2)[c] 2SLS	(3)[c] OLS	(3)[c] 2SLS
Constant	n.a.	−.421[d]	−.463[d]	−.532[d]	−.450[e]	−.629[d]	−.568[d]
		(.083)	(.137)	(.191)	(.240)	(.203)	(.255)
Change in ln(output)	−.619	.670[d]	.637[d]	.679[d]	.769[d]	.670[d]	.732[d]
	(.888)	(.041)	(.085)	(.041)	(.120)	(.042)	(.120)
*Growth	.059	−.627[d]	−.644[e]	−.711[d]	−1.304[d]	−.710[d]	−1.197[d]
	(.222)	(.157)	(.329)	(.168)	(.521)	(.167)	(.510)
West Germany	.311	.451[d]	.479[d]	.520[d]	.478[d]	.556[d]	.529[d]
	(.469)	(.088)	(.114)	(.112)	(.141)	(.114)	(.143)
France	.246	.247[d]	.276[d]	.234[d]	.217[e]	.262[d]	.255[d]
	(.434)	(.089)	(.095)	(.089)	(.110)	(.091)	(.111)
Belgium	.066	.554[d]	.581[d]	.507[d]	.391[e]	.484[d]	.400[d]
	(.250)	(.132)	(.150)	(.138)	(.198)	(.138)	(.190)
Luxembourg	.066	.168	.187	.298[e]	.316	.325[e]	.345[e]
	(.250)	(.130)	(.143)	(.173)	(.194)	(.173)	(.189)

Italy	.098 (.300)	.517[d] (.118)	.553[d] (.146)	.513[d] (.122)	.405[d] (.186)	.532[d] (.122)	.454[d] (.184)
Regional dependence on steel for employment	.738 (.444)	—	—	−.072 (.071)	−.149 (.099)	−.075 (.070)	−.139 (.095)
Regional unemployment	6.25 (2.31)	—	—	.024 (.023)	.040 (.029)	.027 (.023)	.040 (.028)
Change in product wage	.586 (.429)	—	—	—	—	.096 (.072)	.090 (.081)
R^2		.907	.905	.910	.888	.913	.898

a. Dependent variable: change in ln(employment). Its mean (standard deviation) is −.574 (.683). A dash appearing in a column indicates that the variable was not included in the regression equation.

b. Standard deviations are in parentheses.

c. Standard errors are in parentheses. Number of observations is 61. OLS = ordinary least squares model. 2SLS = two-stage least squares model.

d. Coefficient significant at the .05 level or better.

e. Coefficient significant at the .10 level.

Table 5.4. Employment adjustment, 1974–1977, controlling for quality of capital stock.[a]

		(1)[c]		(2)[c]		(3)[c]	
Variable	Mean[b]	OLS	2SLS	OLS	2SLS	OLS	2SLS
Constant	n.a.	.131	.095	.127	.139	.240	.252
		(.129)	(.141)	(.169)	(.173)	(.209)	(.213)
Change in ln(output)	−.314	.455[d]	.401[d]	.337[d]	.297[d]	.327[d]	.338[d]
	(.376)	(.050)	(.142)	(.064)	(.137)	(.073)	(.132)
*Growth	.043	−.407[d]	−.353	−.262[d]	−.157	−.248[e]	−.246
	(.173)	(.110)	(.307)	(.118)	(.267)	(.143)	(.280)
West Germany	.315	.046	.048	.031	.030	.023	.024
	(.468)	(.035)	(.037)	(.034)	(.034)	(.038)	(.038)
France	.247	.067[e]	.073[e]	.051	.048	.025	.024
	(.434)	(.037)	(.038)	(.038)	(.038)	(.045)	(.046)
Belgium	.055	−.030	−.029	−.039	−.041	−.077	−.077
	(.229)	(.062)	(.063)	(.059)	(.060)	(.072)	(.073)
Luxembourg	.068	−.225[d]	−.219[d]	−.277[d]	−.282[d]	−.299[d]	−.298[d]
	(.254)	(.058)	(.059)	(.058)	(.060)	(.065)	(.066)
Italy	.068	.099	.109	.078	.072	.068	.069
	(.254)	(.063)	(.067)	(.070)	(.072)	(.086)	(.087)
Regional dependence on steel for employment	.781	−.019	−.030	—	—	—	—
	(.417)	(.035)	(.039)				

Initial labor productivity	2.791 (2.213)	−.024[d] (.007)	−.023[d] (.008)	−.018[d] (.007)	−.020[d] (.007)	−.033[d] (.013)	−.033[d] (.014)
Year of construction of hot-rolling equipment[f]	55.0 (19.0)	−.079 (.074)	−.007 (.075)	−.118 (.082)	−.117 (.084)	−.121 (.087)	−.124 (.088)
Year of modernization of hot-rolling equipment[f]	67.1 (9.5)	.021 (.158)	.046 (.162)	−.083 (.231)	−.120 (.245)	−.158 (.254)	−.152 (.265)
Year of construction of steel-making equipment[f]	60.3 (16.5)	—	—	.073 (.083)	.077 (.088)	.057 (.089)	.059 (.094)
Diameter of blast furnace (in meters)	8.04 (2.07)	—	—	—	—	−.002 (.008)	−.002 (.009)
R^2		.715	.710	.623	.617	.636	.635

a. Dependent variable: change in ln(employment). Its mean (standard deviation) is −.118 (.186) for columns labeled (1); −.105 (.147) for columns labeled (2); and −.111 (.152) for columns labeled (3). Number of observations is 73 for the columns labeled (1); 63 for the columns labeled (2); and 56 for the columns labeled (3). OLS = ordinary least squares model. 2SLS = two-stage least squares model. A dash appearing in a column indicates that the variable was not included in the regression equation.

b. Standard deviations are in parentheses. Means and standard deviations are for maximum number of observations. They do not vary substantially with the sample size, however.

c. Standard errors are in parentheses.

d. Coefficient significant at the .05 level or better.

e. Coefficient significant at the .10 level.

f. Coefficient and standard error multiplied by 100.

Table 5.5. Employment adjustment, 1974–1982, controlling for quality of capital stock.[a]

Variable	Mean[b]	(1)[c]		(2)[c]		(3)[c]	
		OLS	2SLS	OLS	2SLS	OLS	2SLS
Constant	n.a.	−.898[d]	−1.172[d]	−1.094[d]	−1.170[d]	−.990[d]	−.957[e]
		(.319)	(.436)	(.343)	(.376)	(.389)	(.493)
Change in ln(output)	−.619	.702[d]	.778[d]	.673[d]	.621[d]	.657[d]	.556[d]
	(.888)	(.041)	(.138)	(.040)	(.079)	(.040)	(.084)
*Growth	.059	−.732[d]	−1.338[e]	−.629[d]	−.515	−.679[d]	−.052
	(.222)	(.180)	(.716)	(.156)	(.375)	(.202)	(.584)
West Germany	.311	.513[d]	.485[d]	.454[d]	.499[d]	.504[d]	.620[d]
	(.467)	(.180)	(.136)	(.079)	(.101)	(.081)	(.124)
France	.246	.240[d]	.230[d]	.323[d]	.360[d]	.353[d]	.404[d]
	(.434)	(.088)	(.111)	(.086)	(.095)	(.091)	(.113)
Belgium	.066	.446[d]	.362[e]	.550[d]	.590[d]	.573[d]	.675[d]
	(.250)	(.135)	(.202)	(.117)	(.130)	(.132)	(.168)
Luxembourg	.066	.260	.330	.124	.171	.137	.235
	(.250)	(.173)	(.201)	(.134)	(.146)	(.135)	(.170)
Italy	.098	.552[d]	.517[d]	.664[d]	.713[d]	.686[d]	.800[d]
	(.300)	(.124)	(.168)	(.125)	(.138)	(.134)	(.174)
Regional dependence on steel for employment	.738	−.071	−.131	—	—	—	—
	(.444)	(.068)	(.101)				

Variable							
Regional unemployment	6.25 (2.31)	.030 (.022)	.046 (.031)	—	—	—	—
Initial labor productivity	2.85 (2.35)	−.012 (.015)	.007 (.025)	−.010 (.014)	−.012 (.018)	−.016 (.024)	−.024 (.029)
Year of construction of hot-rolling equipment[f]	55.5 (19.3)	−.175 (.175)	−.115 (.205)	−.317[e] (.186)	−.304 (.195)	−.379[d] (.182)	−.446[e] (.224)
Year of modernization of hot-rolling equipment[f]	67.2 (10.1)	.712[d] (.330)	1.057[e] (.544)	1.479[d] (.452)	1.476[d] (.485)	1.149[d] (.454)	1.053[e] (.536)
Year of construction of steel-making equipment[f]	59.9 (17.5)	—	—	−.244 (.158)	−.232 (.166)	−.261[e] (.154)	−.277 (.180)
Diameter of blast furnace (in meters)	8.21 (2.08)	—	—	—	—	.017 (.016)	.009 (.022)
R^2		.922	.904	.934	.931	.949	.931

a. Dependent variable: change in ln(employment). Its mean (standard deviation) is −.574 (.683) for columns labeled (1); −.530 (.654) for columns labeled (2); and −.563 (.685) for columns labeled (3). Number of observations is 61 for columns labeled (1); 54 for columns labeled (2); and 48 for columns labeled (3). OLS = ordinary least squares model. 2SLS = two-stage least squares model. A dash appearing in a column indicates that the variable was not included in the regression equation.

b. Standard deviations are in parentheses. Means and standard deviations are for the maximum number of observations. They do not vary substantially with sample size, however.

c. Standard errors are in parentheses.

d. Coefficient significant at the .05 level or better.

e. Coefficient significant at the .10 level.

f. Coefficient and standard error multiplied by 100.

results in Tables 5.2 and 5.3 is the large negative sign on the growth dummy interaction variable. The negative coefficient on this variable implies that the elasticity of employment with respect to production is lower for works that are expanding than for works that are contracting. In fact, the negative coefficient on the growth interaction variable is so large as to suggest that there was no expansion—or that there may even have been some contraction—of employment associated with an increase in production. The lower adjustment of employment in expanding works is probably due, in part, to the extreme uncertainty about the industry's future and to the strong job rights, which became explicit or binding during that period. The result supports the popular hypothesis that one of the main effects of strong job security is to discourage new hiring. However, the growth variable may be picking up other factors inadequately controlled for in these regressions. Expanding plants were, on average, more modern and more capital intensive, and therefore may have had a lower employment elasticity for technological reasons.

The results of the second and third sets of equations reported in Tables 5.2 and 5.3 include controls for labor market conditions. Higher regional dependence on steel for employment and a higher unemployment rate are expected to mitigate employment reductions, because they lower the opportunity cost of labor. Only the unemployment variable has the predicted sign, and neither variable is statistically significant in any of the equations. The unemployment figures for 1979, the earliest available figures that are comparable across countries, probably measure unemployment too late to capture regional employment conditions at the beginning of the crisis. Instead, there may be a problem of reverse causality in which higher regional unemployment figures reflect decisions to rapidly reduce steel employment. Because this problem is likely to be particularly serious for the 1974–1977 estimates, the variable is excluded from the remaining equations reported for this time period. Its omission, however, has little effect on the other estimated coefficients.

In the adjustment model of equation (5.1), wages and employment levels are negotiated simultaneously by a union that has a utilitarian objective function. An alternative hypothesis is that contract negotiations are not efficient; that unions and management, at least in certain countries, negotiate only over wages; and that once wage levels are set, management selects employment to lie along its labor demand curve. Unions in different countries also may have different implied objectives, resulting in different trade-offs between wages and employment levels. Unions in some countries may have opted for higher wages and lower employment levels than unions in other countries, all else being constant.

To take account of these factors, equations also were estimated with a control for the change in wage levels. The results of the third set of equations reported in Tables 5.2 and 5.3 include the change in the wage deflated by the product price. In the two-stage least squares estimates, both the wage and the production variables are treated as endogenous. The coefficient on the product wage variable is positive, though insignificant. Because of the possible unreliability of product price data, wages also were deflated by the consumer price index. However, substitution of this real wage measure for the product wage produced similar results, and this variable is excluded from the equations reported below.

The adjustment of work force levels and the observed level of job security were likely to have been heavily influenced by cost pressures to restructure and improve labor productivity during the period. Cross-country differences in employment adjustment in the above equations may be a result of average differences in productivity at the beginning of the adjustment period. Britain and France suffered from old plant and equipment, and, particularly in the case of Britain, from poor labor relations. As a result, these countries had low labor productivity and may have faced greater pressure to modernize and to reduce work force levels.

In Tables 5.4 and 5.5 I attempt to control for these factors by including measures of productivity and measures of the quality and vintage of the capital stock in the beginning of the period in the regression equations.[15] Since some data are missing and since not all works produce iron and steel, the number of observations drops substantially when steel- and iron-making equipment variables are included in the regressions.

Over the 1974–1977 period, low productivity in the beginning of the adjustment period seems to have provided an important stimulus to employment reductions. In Table 5.4 the variable measuring initial productivity is negative and significant in all equations.

Over the 1974–1982 period, the variable that measures the year of last modernization of hot-rolling equipment appears to supplant the one that measures initial labor productivity in explaining employment adjustment. In Table 5.5 the coefficient on this variable is positive and significant in all equations, implying that works with equipment that was more recently modernized reduced employment less. The coefficients on the other variables measuring the quality and vintage of the capital stock generally are insignificant in both tables.

Most important, the inclusion of variables measuring productivity and the vintage of the capital stock does not affect the findings of cross-country differences in the adjustment of employment levels over either period. Over the longer period reported in Table 5.5, the coefficients on the German, French, Belgian, and Italian dummy variables are similar in magnitude to those reported in Table 5.3 and are significant in all equations.

In Table 5.6, the European Community's classification of the hot-rolling and steel-making equipment in the works is used as a measure of the quality and vintage of the capital stock for equations estimated over the entire period. Because the equipment classification variable is measured as of 1982, I did not estimate these equations for the shorter time period. The higher the rating, the greater the technical efficiency, and so the expected sign on these variables is positive. The coefficients on these variables are insignificant, however, and their inclusion does not affect findings concerning the strong differences in employment adjustment between Britain and the Continental countries.

The Adjustment of Employment and Average Hours

Less adjustment of employment levels does not imply less adjustment of total labor input. A plant with strong job security may reduce average hours per worker and still achieve substantial reduction of labor input. Temporary short-time work schemes, increases in vacation time, and permanent reductions in the work week were important mechanisms for avoiding layoffs in Germany, France, Belgium, the Netherlands, and Italy.

In the equations of employment adjustment reported above, country dummy variables are capturing average differences in hours adjustment, productivity growth, or both. Unfortunately, I do not have hours data at the plant level to test this trade-off between employment and hours directly. In the absence of plant-level hours data, various pieces of information on hours at the country level are used to bring some evidence to bear on the extent to which hours reduction was a substitute for work force reduction during restructuring.

Table 5.6. Employment adjustment, 1974–1982, controlling for technical efficiency.[a]

		(1)[c]		(2)[c]	
Variable	Mean[b]	OLS	2SLS	OLS	2SLS
Constant	n.a.	−.335[d] (.118)	−.424[d] (.166)	−.395[d] (.144)	−.517[d] (.246)
Change in ln(output)	−.631 (.901)	.665[d] (.039)	.594[d] (.082)	.597[d] (.082)	.461[d] (.200)
*Growth	.062 (.227)	−.634[d] (.173)	−.416 (.397)	−.479[d] (.223)	−.073 (.696)
West Germany	.310 (.467)	.446[d] (.085)	.508[d] (.114)	.460[d] (.105)	.501[d] (.128)
France	.224 (.421)	.310[d] (.086)	.339[d] (.093)	.386[d] (.117)	.420[d] (.132)
Belgium	.069 (.256)	.519[d] (.126)	.576[d] (.142)	.530[d] (.139)	.545[d] (.149)
Luxembourg	.069 (.256)	.160 (.127)	.208 (.139)	.081 (.161)	.080 (.176)
Italy	.103 (.307)	.523[d] (.112)	.599[d] (.138)	.548[d] (.153)	.571[d] (.165)
Initial labor productivity	2.75 (2.28)	−.012 (.017)	−.015 (.023)	−.026 (.020)	−.043 (.039)
Classification of hot-rolling equipment	3.28 (1.24)	−.015 (.031)	−.016 (.037)	−.020 (.038)	−.001 (.053)
Classification of steel-making equipment	3.00 (1.46)	—	—	.019 (.032)	.020 (.038)
R^2		.924	.918	.781	.754

a. Dependent variable: change in ln(employment). Its mean (standard deviation) is −.568 (.683) for columns labeled (1) and −.362 (.402) for columns labeled (2). A dash appearing in a column indicates that the variable was not included in the regression equation.

b. Standard deviations are in parentheses. Means and standard deviations are for maximum number of observations. They do not vary substantially with sample size, however.

c. Standard errors are in parentheses. Number of observations is 58 for columns labeled (1) and 44 for columns labeled (2). OLS = ordinary least squares model. 2SLS = two-stage least squares model.

d. Coefficient significant at the .05 level or better.

e. Coefficient significant at the .10 level.

Figure 5.1 shows average annual hours worked by manual workers in each country from 1974 to 1983.[16] In 1974 average hours were similar across countries. In countries using short-time work (Germany, France, Belgium, and Italy), average hours dropped sharply after 1974 and fluctuated considerably thereafter. In Luxembourg they also dropped, though less sharply; this decline is largely accounted for by the elimination of overtime work and by the increase in paid vacation. In Britain average hours were relatively stable, except in 1980, which was a strike year. In 1983 average hours in Britain were 16 percent greater than those in Luxembourg, 19 percent greater than those in France, 23 percent greater than those in Germany and Belgium, and 31 percent greater than those in Italy.[17]

Table 5.7 shows the percent change in crude steel production, manual worker employment, and manual worker hours over the 1974–1982 period for five of the countries in the study.[18] The difference between the percent change in employment levels and the percent change in total hours worked is the greatest in countries that extensively used short-time work: Germany, France, and Belgium. The percent changes in employment and total hours are divided by the percent change in production to yield crude employment- and hours-output elasticities.

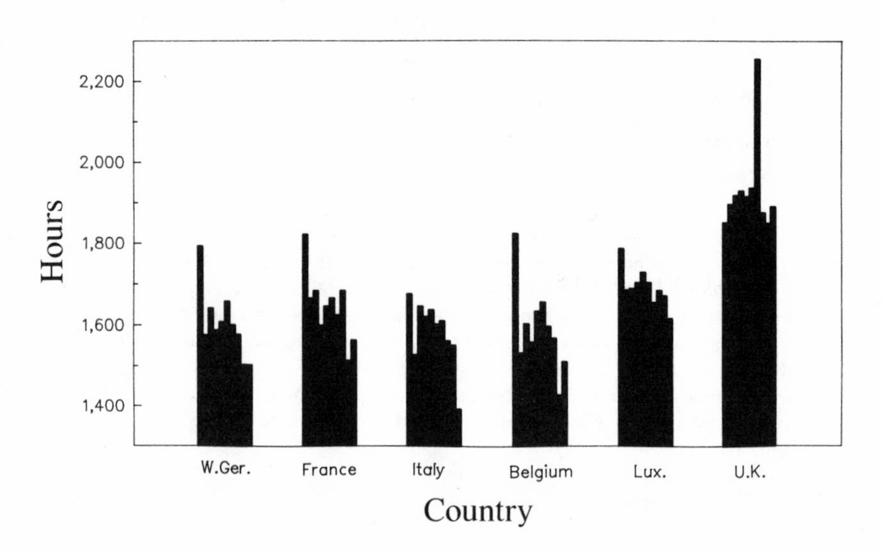

Figure 5.1. Hours worked per manual worker, 1974–1983. (*Source:* Author's calculations based on EUROSTAT data.)

Table 5.7. Percent changes in crude steel production, manual workers and manual worker hours, 1974–1982.[a]

	Percent change in output	Percent change in employment	Percent change in total hours	Employment-output elasticity	Hours-output elasticity
West Germany	−32.6	−24.4	−36.7	0.75	1.13
France	−31.9	−46.3	−52.5	1.45	1.65
Belgium	−39.0	−34.6	−48.8	0.89	1.25
Luxembourg	−45.6	−50.6	−53.9	1.11	1.18
U.K.	−38.6	−59.9	−61.4	1.55	1.59

a. Table gives author's calculations based on EUROSTAT data. Production data are from *Iron and Steel Yearbook*, 1983. Employment and hours data are from *Emploi et Chomage*, September 30, 1983.

The employment elasticities conform with the patterns observed in the regressions using plant-level data. The employment elasticity for Britain is the greatest, followed by those for France and Luxembourg, all of which are greater than 1. The employment elasticities for Belgium and Germany are both less than 1.

The hours elasticities present a quite different picture, however. Because average hours in Britain remained at about the same level over the period, there is little difference between Britain's employment and hours elasticities. In contrast, the hours elasticity is substantially higher than the employment elasticity in the Continental countries, particularly Germany, Belgium, and France, which used short-time work schemes. France's hours-output elasticity is actually the highest, and the gap between Britain and the other Continental countries is substantially smaller when one compares the hours rather than the employment elasticities.

Comparisons across countries of these crude employment- and hours-output elasticities should be used only as a rough indicator of the importance of average hours adjustment. In particular, the differential between total hours elasticities cannot be ascribed to productivity differences. The numbers have not been adjusted for output mix or for the degree of subcontracting, among other things. In countries with strong job security, some companies took over maintenance and construction work in order to preserve jobs for steelworkers, whereas in Britian there reportedly was greater outsourcing of these tasks to lower-cost subcontractors.[19] Without appropriate adjustments, differentials may be incorrectly interpreted as differences in productivity growth. In Chapter 6 we shall look at some comparisons of productivity growth across countries that control for these factors.[20]

Still, these aggregate numbers suggest that a substantial part of the differential in the adjustment of employment levels across countries is accounted for by differences in the adjustment of average hours per worker. Yet how companies adjust labor input—whether through employment levels or average hours worked—has potentially important implications for the allocation of labor in the economy at large. In general, one would expect that laid-off workers would face greater economic incentives to find new employment than workers on short-time work. It is likely that relatively few steelworkers on short-time found second jobs to make up for their shortfall in hours, particularly since they received unemployment insurance from the state and, in many cases, supplemental benefits from their companies. Therefore, one might expect greater utilization of labor overall in Britain than in Continental countries that extensively used short-time work.

The opposite may be true, however, if geographic mobility is low and rapid work force reductions overwhelm a local economy. In such cases, rapid work force reductions may result in high and persistent unemployment. Allegations of this kind were made in the case of Britain, in particular. One follow-up study of displaced workers in Wales showed very high rates of unemployment. More than two years after layoff, of those who were still in the labor force, 45 percent were unemployed (Harris 1984).

Luxembourg combined strong job guarantees with high utilization of its labor force. All workers were guaranteed their income. Although the state and the European Community covered about 15 percent of the wages for workers in the Anticrisis Division, the company still had an economic incentive to utilize this labor in its highest-paying alternative. If companies are better at job search activities than individuals are, such a subcontracting arrangement may be efficient. The structure of Luxembourg's work force reduction program, therefore, may have resulted in more efficient utilization of labor than was achieved in other Continental countries with similar levels of job guarantee.[21] However, the 2 percent unemployment rate in Luxembourg was not comparable to the double-digit unemployment rates experienced in other European steel regions during restructuring. If regional unemployment is already high and the opportunity cost of labor low, the social insurance scheme for short-time work may introduce little or no distortion in the allocation of labor, and thus may represent a relatively efficient means of work force reduction, given the level of job guarantee.

Conclusion

Differences in job security appear to explain large and statistically significant cross-country differences in the adjustment of employment levels in the E.C. steel industry. In the years immediately following the downturn, job security was relatively strong and work force reduction practices were similar in all countries. Correspondingly, over the 1974–1977 period, the adjustment of employment levels to changes in production was similar in all countries. Companies tended to view the crisis as primarily cyclical prior to 1977, which may explain the fact that most works adjusted their employment levels relatively little during the period. The exception is Luxembourg. With a third of its industrial labor force employed in steel in 1974, the government, in cooperation with management and labor, acted quickly to move workers out of that sector, at least temporarily. In these early years, much of ARBED's excess labor was reassigned to public works projects.

By the 1980s the crisis was widely recognized as structural, and substantial differences in the level of job security and approaches to work force reductions had arisen between Britain and the Continental countries. Over the 1974–1982 period, plants in Britain reduced employment significantly more than in Germany, France, Belgium, and Italy. These differences cannot be ascribed to differences in productivity or the quality of the capital stock that may have placed greater cost pressures on Britain to reduce work force levels. The regression equations show that although these factors were important determinants of work force reductions, they do not explain the large cross-country differences in employment adjustment.

Employment adjustment in Luxembourg was not significantly different from that in Britain. The rapid reduction of work force levels implied by these results, however, is accounted for by ARBED's Anticrisis Division. France is a borderline case. Although the empirical analysis indicates that employment adjustment was significantly lower there than in Britain, it also was significantly greater than in other Continental countries. France achieved quite rapid work force reductions, despite strong job security during this period, through costly early retirement and job buyout programs. The costs to the steel companies and the government of strong job guarantees did result, however, in a weakening of these guarantees in 1984.

Underlying some of the differences across countries in the adjustment of employment levels are differences in the adjustment of average hours per worker. In Germany, France, Belgium, Italy, and to a lesser extent Luxembourg, companies relied on some combination of schemes to reduce working time (increased vacation time, reduction in the average work week, and temporary short-time work) to compensate for excess employment. In contrast, in Britain average production worker hours remained about the same during restructuring.

A strategy of utilizing reductions in average hours to avoid layoffs may have important implications for employment in the economy. Even qualitatively, however, its effects are difficult to predict, and they depend, in part, on the dynamics of the regional economies, which often were dominated by steel. On the one hand, the more quickly workers are laid off from the steel sector, the more quickly they are likely to be employed in growing sectors. On the other hand, a policy that spreads out work force reductions over time may allow a regional economy to better adapt to structural change and may result in less overall unemployment.

The balance struck between the adjustment of average hours and employment levels also may have important implications for the allocation of capital. The cost of using measures to reduce work time to avoid layoffs at a works is likely to grow at an increasing rate as production declines. With strong job security, total plant closures necessitate costly, often unpopular transfers. Therefore, in countries where there was a tendency within plants to spread available work over the number of workers, there also may have been a tendency to spread production quotas and capacity reductions over plants. The broader implications of job security on the allocation of production, investment, and productivity growth are explored next.

6

Job Security and Restructuring: The Allocation of Production and Capacity in Steel

Prior to the crisis, much of the European steel industry had become less competitive in world markets, owing to its age and to major process innovations introduced after World War II. Although restructuring had begun before the crisis, the sharp drop in demand in 1975 accelerated the process. The widely advocated strategy for rationalizing the industry consisted of closing the older, smaller works that were no longer optimally located; concentrating production at newer, larger sites that often were located on the coast, in order to take advantage of economies of scale and lower transportation costs; and modernizing equipment at works that remained open.

Strong job security, coupled with depressed regional economies and low labor mobility, rendered this strategy costly, however. These labor market factors would be expected to result in less concentration of production, so as to preserve traditional patterns of employment, and consequently would lead to more modernization of inefficient equipment at suboptimally located works.

The preceding chapter examined the effects of job security on a plant's adjustment of employment levels, given some change in production. But strong job security also may affect employment through its impact on production, investment, and closure decisions.

The Determinants of the Allocation of Production and Capacity: The Estimating Framework

The major integrated steel producers, with the support of the European Commission, began negotiating production quotas and product prices for key hot-rolled product lines in 1977. In addition, investments and capacity reductions at the hot-rolling stage were negotiated on a company-by-company basis within the European Community. The initial goal was to bring capacity in line with production by 1986. Therefore, the allocation of production throughout most of the adjustment period and the allocation of capacity in the medium run approximated a multilateral negotiated solution to an industrial restructuring problem.

The cost minimization problem facing individual firms analyzed in the preceding chapter essentially is embedded in a larger problem of resource allocation within the Community. With a cooperative solution, and in the absence of transactions costs to side payments, the cost minimization problem in equation (5.1) may be easily extended to incorporate the allocation of production and capacity across plants over time. Steel producers, acting in cooperation with the European Commission, optimally would seek to allocate production quotas so as to minimize costs, subject to constraints on worker utility or job rights, which may vary over time and across countries. Formally, the problem may be written

$$\min_{q,L,I,s,w_1} \quad \sum_t \sum_i [w_{1it} L_{it} + s_{it}(L_{i,t\text{-}1} - L_{it}) + r_t I_{it} + m_{it} q_{it}] \; (\delta)^{-t} \tag{6.1}$$

subject to

$$\frac{L_{it}}{L_{i,t\text{-}1}} v(w_1) + \frac{L_{i,t\text{-}1} - L_{it}}{L_{i,t\text{-}1}} [v(w_2 + s) - B] \geq V_{it}, \text{ for all } i,t$$

$$\sum_i q_{it} = Q_t, \text{ for all } t,$$

where δ is the discount factor, $q_t = q(L_t, K_{vt}(K_{vt-1}, I_t))$, and m represents unit transportation costs. The notation t indexes time, and i indexes the plant. All other variables are defined as in equation (5.1).

The only new cost factor introduced in equation (6.1) is transportation costs. Reflecting the changing relative costs of raw materials and transportation by water, the optimal location of integrated steel mills changed from inland sites near local raw materials to coastal sites, which enjoy cheap transportation costs for imports of raw materials and for exports of finished products. This locational cost factor is captured in the model simply as a unit cost of output that varies across plants.[1]

The allocation of production and capacity across plants is an implicit function of the exogenous variables in (6.1). These include past levels of production or capacity, the initial vintage of the capital stock, regional labor market conditions, transportation costs, and job rights. The empirical analysis in this chapter focuses on whether worker rights to job security affected the allocation of production and capacity in the E.C. steel industry. As above, the level of job rights is taken to be similar across plants within countries, though these rights may change over time.

The null hypothesis is that observed differences in job security across countries do not result from underlying differences in job rights and have no causal effect on allocation decisions. Rather, causality runs in the other direction. Workers in countries with less efficient plant infrastructure and equipment, for example, may suffer greater production and capacity reductions, resulting in weaker job security.

Any effect that job security has on production and capacity allocation may be evidenced in several ways, given the available data. To the extent that job security is generally strong in Community countries, there is likely to be some equalization of cutbacks in production and capacity across companies. Spreading production and capacity reductions across companies in the Community facilitates the use of alternatives to layoffs to achieve work force reductions.

To the extent that job security differs across countries, one would expect the relative importance of the determinants of production and capacity allocation to differ. In cross-section data, technological and locational cost factors, for example, would be expected to carry less weight in allocation decisions where job rights are stronger.

Job security changed over time in certain countries. Therefore, in time series data, one would expect changes in the relative importance of the allocation criteria that are consistent with changes in job security. A weakening of job rights should be accompanied by a shift in the allocation of production toward plants with technological and locational cost advantages.

In addition, there may be substantially more rationalization at the product level than at the works level. Employment considerations inhibit plant closures. However, for any given level of hot-rolled capacity, there may be considerable scope for improving efficiency through the concentration of production at the product level. Trades of capacity across product lines between plants make possible greater exploitation of economies of scale in production at the hot-rolling stage and utilization of the most efficient hot-rolling equipment.

Finally, technical and locational factors should play a greater role in allocation decisions over time. Time mitigates the employment adjustment problem, because of the natural attrition of the work force and because jobs are created in other industries to replace the losses in steel.

The Allocation of Production, 1975–1982

The E.C. steel industry, as a whole, lagged behind the most efficient world producers. There was, however, considerable variability within the Community in the age and technical efficiency of the capital stock. Table 6.1 summarizes measures of the vintage of the capital stock as of 1977 by country and for the European Community as a whole for the works in the sample. Weighted and unweighted means and standard deviations are provided for the year of construction of the plant's equipment at the hot-rolling, steel-making, and iron-making stages; for the year of last modernization at the hot-rolling and steel-making stages; and for the diameter of the blast furnace.

For fully integrated plants, one of the best indicators of the age and technical efficiency of the works overall is the diameter of the blast furnace. The diameter of the blast furnace was relatively small in Luxembourg, Belgium, and Britain.[2] For integrated works, capacity was also negatively correlated with the age of the works. The competitive disadvantage that many integrated works in Britain faced because of older plant infrastructure and equipment is particularly apparent in their small average size, which is shown below in Table 6.6.

Although there were differences in the average age of the capital stock across countries, what is perhaps most striking in Table 6.1 is the variation in the age of the equipment in most countries. This fact indicates that there was considerable scope for consolidating production at newer facilities throughout the Community.

After 1977, steel companies, under the direction of the E.C. Commission, negotiated production quotas and minimum prices for certain products. Company production quotas were set by a complicated formula based on historical production levels. The steel companies could trade these production rights. Although trades of production quotas in theory could be utilized to effect large changes in production shares across companies within the Community, the E.C. plan was widely perceived as helping to stabilize shares during restructuring.

Table 6.1. Summary of capital stock measures by country.[a]

Country	Year of construction at hot-rolling stage	Year of modernization at hot-rolling stage	Year of construction at steel-making stage	Year of modernization at steel-making stage	Year of construction at iron-making stage	Diameter of blast furnace (in meters)
West Germany						
Unweighted average	1958	1968	1957	1971	1951	7.7
	(17.6)	(9.0)	(17.2)	(5.8)	(24.9)	(2.6)
Weighted average	1962	1971	1957	1972	1962	9.3
	(11.2)	(6.0)	(19.8)	(5.2)	(12.5)	(1.8)
Belgium						
Unweighted average	1965	1973	1968	1972	1962	7.8
	(3.7)	(2.4)	(4.6)	(3.7)	(3.7)	(2.2)
Weighted average	1964	1973	1967	1972	1961	8.5
	(2.3)	(1.8)	(2.7)	(3.3)	(3.4)	(1.7)
France						
Unweighted average	1952	1964	1960	1973	1953	8.6
	(19.7)	(15.3)	(21.9)	(4.8)	(15.7)	(1.6)
Weighted average	1958	1970	1962	1973	1963	9.1
	(13.3)	(6.6)	(20.3)	(3.6)	(11.7)	(1.5)
Italy						
Unweighted average	1950	1958	1963	1964	1962	8.7
	(22.6)	(14.0)	(8.4)	(7.6)	(8.4)	(1.6)

Weighted average	1967	1967	1969	1969	1968	9.9
	(10.3)	(8.8)	(7.2)	(6.3)	(5.2)	(1.1)
Luxembourg						
Unweighted average	1939	1969	1967	1971	1961	7.1
	(28.1)	(8.5)	(4.8)	(4.9)	(17.0)	(1.8)
Weighted average	1942	1968	1968	1972	1963	7.3
	(25.7)	(7.7)	(4.2)	(4.2)	(12.5)	(1.5)
Netherlands						
Unweighted average	1969	1969	1968	1976	1967	11.0
Weighted average	—	—	—	—	—	—
U.K.						
Unweighted average	1954	1967	1964	1968	1953	7.9
	(21.3)	(6.3)	(8.9)	(4.8)	(11.5)	(1.8)
Weighted average	1959	1970	1966	1970	1955	8.6
	(17.4)	(5.1)	(6.7)	(3.7)	(10.5)	(1.6)
E.C. total						
Unweighted average	1955	1967	1961	1970	1955	8.0
	(19.7)	(10.7)	(15.4)	(5.6)	(18.1)	(2.1)
Weighted average	1961	1970	1963	1972	1962	9.2
	(13.7)	(6.3)	(16.0)	(4.8)	(10.9)	(1.7)

a. Numbers are based on the sample of works in the author's data base. Standard deviations are in parentheses. The weights for the weighted averages are the plant's capacity for the relevant stage of production in 1977.

As evidence of this, Table 6.2 shows each country's share of total hot-rolled production in the European Community from 1974 to 1982. Table 6.3 presents simple correlations between 1974 production shares and production shares for the other years for the eighty-three works in my sample for which data on production were complete. These correlations are calculated at both the plant and the company level, for total hot-rolling as well as for individual product lines.

Market shares remained fairly stable over the period, despite the volatility of prices and production. To put these figures in some perspective, a high-level E.C. official remarked during an interview that had free market forces been allowed to operate, the German industry probably would have increased its market share from 33 percent to 50 percent and the Italian from about 15 percent to 25 percent. Such a claim reveals what people close to the industry and to the negotiations thought was the large potential for change in the structure of production within the European Community. Italy did increase its market share substantially, and Britain did not fully regain the market share it lost following the 1980 strike. Still, the large shifts thought possible by some analysts did not occur.

The simple correlations in Table 6.3 also generally support the view that the Davignon Plan helped stabilize market shares and thus preserve the status quo. Particularly for total hot-rolled products at the company level, the correlations remained quite high in the 1970s, though they fell in the 1980s during the deep recession. Changes in market shares are more evident for certain product groups. Although greater rationalization at the product level is expected, it may have been due to factors largely beyond the control of the Davignon Plan.

Table 6.2. Shares of E.C. hot-rolled production, 1974–1982 (in percent).[a]

Country	1974	1975	1976	1977	1978	1979	1980	1981	1982
West Germany	32.7	31.1	30.9	29.5	29.5	30.6	32.1	31.5	31.2
Belgium	10.7	8.8	9.2	9.5	10.3	10.4	10.4	10.0	9.3
France	17.7	17.0	18.5	18.3	18.7	17.9	19.3	18.0	17.9
Italy	16.0	21.0	17.7	19.4	18.7	18.3	21.1	20.6	21.9
Luxembourg	3.9	3.3	3.3	3.4	3.4	3.4	3.7	3.1	3.2
Netherlands	4.1	3.4	4.1	4.0	4.2	4.1	3.8	4.0	3.9
U.K.	14.5	14.9	15.7	15.3	14.5	14.5	9.0	12.1	12.1

a. For these years the European Community includes the seven countries in this study plus Denmark and Ireland. Data are author's calculations based on a complete sample of plants.

Table 6.3. Correlations of production share in 1974 with that in other years.[a]

Product	Number of observations	1975	1976	1977	1978	1979	1980	1981	1982
Total hot-rolling									
Company	28	.985	.985	.979	.983	.989	.940	.953	.930
Plant	83	.971	.966	.953	.956	.964	.935	.896	.872
Plate									
Company	17	.985	.958	.953	.957	.926	.941	.953	.896
Plant	29	.980	.950	.948	.952	.883	.956	.954	.909
Hot-wide strip									
Company	15	.974	.984	.954	.952	.972	.905	.924	.882
Plant	20	.969	.973	.930	.920	.942	.897	.878	.847
Sections									
Company	19	.994	.990	.972	.969	.973	.925	.767	.763
Plant	41	.981	.966	.957	.958	.946	.915	.529	.508
Bars and wire rod									
Company	22	.952	.937	.908	.947	.967	.872	.819	.819
Plant	60	.883	.869	.811	.875	.887	.823	.670	.677
Hoop and strip									
Company	9	.948	.945	.943	.966	.962	.929	.860	.864
Plant	15	.908	.909	.906	.907	.908	.428	.439	.392

a. Based on a sample of 83 works in the author's data base.

For example, because of process innovations, for most purposes hoop and strip can be more cheaply produced now by cutting hot-wide strip; consequently these mills were being closed for technological reasons. Also, the market for bars was particularly difficult for the E.C. Commission to control because of the number of small producers; the major integrated producers of EUROFER generally were losing market share in this product line to mini-mills.

Even if the Davignon Plan effectively stabilized production shares at the country or company level, there was considerable scope for reallocating production across plants within countries or companies. One would expect that technical factors would be more important determinants in allocating production across plants in countries where the adjustment of employment is relatively great or job security is relatively weak. One also would predict that the determinants of production allocation within countries would change if workers' rights to job security changed.

Cross-section and time series data were pooled to formally test for the determinants of the allocation of production over the 1975–1982 period within the Community. The key hypotheses to be tested are, first, whether in countries with weaker job security, particularly Britain, technological and other cost factors were more important determinants of the allocation of production than in countries with stronger job security; and, second, whether in Britain, where job security was significantly weakened in 1980, technological and other cost factors became more important determinants of production allocation following the policy change.

Based on the resource allocation problem in equation (6.1), the following equation was estimated:

$$
\begin{aligned}
(6.2) \qquad q_{it} = {} & \beta_0 + \beta_1 q_{i74} + \sum_j \beta_{2j} X_{ij} + \sum_j \beta_{3j} X_{ij} t + \sum_k \sum_j \beta_{4jk} X_{ij} C_k \\
& + \sum_j \beta_{5j} X_{ij} UK_{80} + \sum_l \beta_{6l} D_{il} + \sum_l \beta_{7l} D_{il} t + \\
& + \sum_t \beta_8 q_{i74} \mathrm{Year}_t + \sum_t \sum_k \beta_{9kt} q_{i74} C_k \mathrm{Year}_t + \varepsilon_{it}.
\end{aligned}
$$

The dependent variable, q_{it}, is the share of production of plant i in year t; it is a function of the plant's initial production share (1974 is taken as the base year) and a number of other variables, all interacted with the initial production share. The term q_{i74} is the plant's share of production in 1974; X_j are other exogenous variables in (6.1), including the initial vintage of the capital stock and plant location, interacted with initial production share; D_l are product mix variables, defined as the initial share of a plant's production in various product lines interacted with q_{i74}; C_k are country dummy variables; UK_{80} is a dummy variable for Britain for the years beginning in 1980; and $Year_t$ denotes year dummy variables.[3]

The right-hand variables X_j include measures of the vintage of the capital stock at the hot-rolling, steel-making, and iron-making stages. To capture transportation cost differences, a dummy variable equal to 1 if the works has a coastal location, interacted with initial production share, is included in the equations.[4]

The specification in equation (6.2) allows the effects of selected exogenous variables to vary over time and across countries. One would expect that technical and locational factors would have become increasingly important determinants of production over time, as the structural nature of the excess capacity became increasingly accepted. The strength of the job guarantees of workers varied across countries, affecting company costs of reducing work force levels. Thus, one might expect other factors to be less important in explaining production allocation in countries with relatively strong job security.

In addition, the strength of the job guarantees that were granted steelworkers did change over time in certain cases. By far the most important policy change occurred in Britain beginning in 1980. The equation allows for a possible shift in the importance of key determinants in Britain.[5]

Product mix variables are included to capture varying demand conditions of different product lines. The equation also includes the interaction of a time trend with the product mix variables. Production shares decreased in the plants in the sample over time, particularly in two product lines: hoop and strip, and bars and wire rod. In the case of the former, technological advances rendered hoop-and-strip mills less efficient than other processes for making this product. In the case of the latter, plants in the sample lost market share to mini-mills.

Finally, the interaction of q_{i74} with country dummy variables and year dummy variables allows average production shares to vary each year across countries.[6] Because of the large number of country interaction terms included in these equations, I report results of equations that include only countries with a large number of plants: Germany, France, and Britain.[7]

Columns (1), (2), and (3) of Table 6.4 report the results of equations

Table 6.4. The determinants of production allocation.[a]

Variable	Mean for column (1)[b]	(1)[c]	Mean for column (2)[b]	(2)[c]	Mean for column (3)[b]	(3)[c]
Share of hot-rolled production, 1974	.011 (.013)	.648[d] (.213)	.011 (.013)	.788[d] (.226)	.013 (.014)	1.414[d] (.316)
Year hot-rolled construction	.645 (.810)	−.223[d] (.055)	.690 (.842)	−.187[d] (.061)	.767 (.885)	−.202 (.067)
*Time[e]	50.6 (63.6)	.281[d] (.070)	54.2 (66.2)	.235[d] (.078)	60.2 (69.5)	.257[d] (.086)
*France	.162 (.415)	.032[d] (.003)	.156 (.427)	.038[d] (.004)	.180 (.461)	.036[d] (.004)
*U.K.	.123 (.278)	.021[d] (.006)	.131 (.292)	.026[d] (.006)	.118 (.271)	.015[d] (.007)
*U.K. year ≥ 1980	.046 (.180)	−.005 (.008)	.049 (.190)	−.005 (.009)	.044 (.176)	−.007 (.011)
Percent steel continuously cast[e]	—	—	.386 (.743)	.481 (1.720)	.442 (.794)	1.120 (1.800)
*Time[e]	—	—	30.3 (58.3)	−.006 (.022)	34.7 (62.4)	−.014 (.023)
*France[e]	—	—	.049 (.284)	.067 (.125)	.054 (.308)	−.051 (.141)
*U.K.[e]	—	—	.050 (.171)	−.739[d] (.155)	.050 (.175)	−.681[d] (.172)
*U.K. year ≥ 1980[e]	—	—	.019 (.108)	.662[d] (.240)	.019 (.110)	.881[d] (.268)
Diameter of blast furnace	—	—	—	—	.115 (.145)	.743 (.429)
*Time	—	—	—	—	9.04 (11.41)	−.010 (.005)
*France	—	—	—	—	.028 (.074)	.071[d] (.027)
*U.K.	—	—	—	—	.017 (.040)	.090[d] (.044)

*U.K. year ≥ 1980	—	—	—	—	.006	.163[d]
					(.026)	(.070)
Coastal plant	.002	−4.464[d]	.002	−5.308[d]	.002	−6.314[d]
	(.006)	(1.083)	(.006)	(1.119)	(.007)	(1.270)
*Time	.140	.058[d]	.157	.068[d]	.186	.082[d]
	(.468)	(.014)	(.493)	(.014)	(.532)	(.016)
Share plate	.001	−.612	.001	.442	.002	1.199
	(.003)	(1.753)	(.003)	(1.846)	(.003)	(2.160)
*Time	.106	.004	.108	−.010	.119	−.021
	(.225)	(.022)	(.232)	(.024)	(.245)	(.027)
Share sections	.001	−3.182	.001	1.661	.001	.957
	(.002)	(2.969)	(.002)	(3.457)	(.002)	(3.981)
*Time	.097	.043	.094	−.019	.096	−.010
	(.159)	(.038)	(.155)	(.044)	(.159)	(.051)
Share bars and wire rod	.003	5.233[d]	.003	4.274[d]	.003	6.158[d]
	(.004)	(1.318)	(.004)	(1.390)	(.004)	(1.764)
*Time	.237	−.075[d]	.249	−.063[d]	.265	−.088[d]
	(.299)	(.017)	(.311)	(.018)	(.331)	(.022)
Share hoop and strip	.0005	6.382[d]	.0004	7.917[d]	.0004	10.043[d]
	(.0015)	(3.333)	(.0015)	(3.587)	(.0016)	(3.932)
*Time	.037	−.085[d]	.033	−.105[d]	.033	−.132[d]
	(.118)	(.042)	(.121)	(.046)	(.126)	(.050)
R^2		.956		.960		.964

a. Dependent variable: share of hot-rolled production. Its mean (standard deviation) is .008 (.012) for the column labeled (1); .009 (.012) for the column labeled (2); and .010 (.013) for the column labeled (3). A constant term and initial capacity share interactions with country and year dummy variables were included in the equations but are not reported. The plant's production share in 1974 is interacted with all of the other right-hand variables in these equations.

b. Standard deviations are in parentheses.

c. Standard errors are in parentheses. Number of observations is 512 for column (1); 456 for column (2); and 384 for column (3). A dash appearing in the column indicates that the variable was not included in the regression equation.

d. Coefficients significant at the .05 level of confidence or better.

e. Coefficients and standard errors are multiplied by 100.

that include equipment variables at the hot-rolling stage, at the hot-rolling and steel-making stages, and at the hot-rolling, steel-making, and iron-making stages, respectively. The number of observations drops across equations because not all plants in the sample were fully integrated and because missing values were a problem, particularly for the variable measuring diameter of the blast furnace.[8]

The interactions of a time trend with a number of variables are included in all equations to capture changes in allocation patterns over time. These time trend interaction variables show significant changes in how production was allocated within the Community over the period. The coefficients on the year of construction of hot-rolling equipment and on the coastal trend variables are positive and significant, indicating that, for the Community overall, production increasingly was allocated to newer, coastal works. In addition, the coefficient on the time interaction with a plant's share in bars and wire rod and in hoop and strip is, as expected, negative, reflecting worsening demand conditions in those product lines.

The equations allow for policy changes in Britain in 1980 that affected job security. Although substantial restructuring took place in Britain throughout the 1970s, much of it focused on plants targeted for closure prior to the crisis; these works are excluded from this study. Restructuring accelerated under Prime Minister Thatcher, who was elected on a platform calling for the reduction of government involvement in the economy. An integral component of this strategy was the end of subsidization to and reprivatization of nationalized industries. Immediately following the largely unsuccessful strike in early 1980, the BSC launched a major program of plant closures and employment reduction.

The impact of this policy change on the allocation of production shows up quite strikingly in the regression equations. For two out of the three equipment variables—the percent of continuously cast steel and the diameter of the blast furnace—the coefficient on the interaction with a dummy variable for Britain during the 1980s is positive and highly significant. These results capture a shift in production toward more modern facilities. The results from these equations also indicate that technical factors were generally more important in allocating production in Britain than in Germany in the 1980s.[9] This finding is consistent with the weaker job security in Britain than in Germany.

The results in Table 6.4 suggest that technical factors were likewise more important in the allocation of production in France than in Germany. The coefficients on the interaction of the dummy variable for France with the hot-rolling equipment variable and with the variable for the diameter of the blast furnace are positive and significant. Although job security in France was strong during this period, the steel industry nonetheless achieved large work force reductions. The relatively large work force adjustment and the relative importance of technical variables in allocating production are, no doubt, part of the same phenomenon.

Trends in labor productivity in the steel industry are a function both of how labor is adjusted for any given reduction in output at the plant level (discussed in the preceding chapter) and of how production is allocated across plants. In Table 6.5, I examine the net impact of these two processes by looking directly at productivity estimates for Germany, France, and Britain from 1972 to 1982. Cross-country comparisons of productivity in the steel industry are notoriously unreliable, because of the difficulty in calculating comparable series and because of the political sensitivity of the numbers. The numbers in Table 6.5 are part of a series constructed by the U.S. Bureau of Labor Statistics.

Table 6.5. Trends in productivity (in tons per person-hour).[a]

Year	France	West Germany	U.K.
1972	59.03	67.84	49.94
1973	62.73	75.52	53.19
1974	65.25	80.44	48.94
1975	56.18	72.14	42.22
1976	61.11	76.68	46.46
1977	64.80	76.83	45.15
1978	72.12	84.67	48.08
1979	79.77	93.12	51.83
1980	86.38	92.35	43.61
1981	88.84	94.73	57.02
1982	83.53	91.20	60.28
Growth rates (in percent)			
1972–1982	3.5	3.0	1.9
1975–1982	5.7	3.3	5.1
1975–1979	8.8	6.4	5.1
1979–1982	1.5	−0.7	5.0

a. Table gives author's calculations based on unpublished data from the U.S. Bureau of Labor Statistics. Productivity numbers represent minimum estimates.

The numbers are adjusted for differences in product mix and for differences in the degree of subcontracting, though not for capacity utilization.

With these caveats in mind, the statistics show quite large differences in productivity and productivity growth across countries. Productivity is the highest in Germany and the lowest in Britain throughout the period. In 1974, labor productivity was more than 60 percent higher in Germany than in Britain, and more than 20 percent higher in Germany than in France. Despite lower productivity, unit labor costs were lower in Britain and France than in Germany, as a result of lower wages. The rate of productivity growth from 1975 to 1982 (both recession years) was 5.7 percent in France, 5.1 percent in Britain, and 3.3 percent in Germany. The patterns of productivity growth are also quite different across countries during subperiods. From 1975 to 1979, which represents the movement from a trough to a peak, the rate of productivity growth was the greatest in France at 8.8 percent, followed by Germany at 6.4 percent, and by Britain at 5.1 percent. However, from 1979 to 1982, which represents the movement from a peak to a trough, Britain maintained its productivity growth, while productivity growth in France and Germany slowed or became negative.

These patterns of productivity growth are consistent with job security in these countries. The spurt of productivity growth in Britain in the 1980s, even as capacity utilization was declining, coincided with massive layoffs. Productivity was more sensitive to the 1982 recession in Germany and France, where layoffs were minimized.

Still, the size of the productivity increases and accompanying work force reductions in France are surprising, given the strong job security there. The work force reductions, attained in large part through early retirement of workers age fifty and over, entailed high costs to the steel companies and the government. Over time, employment reductions and productivity growth became constrained by the age structure of the work force. By 1983, the percentage of the work force in the French steel industry between the ages fifty and fifty-nine was 13.4 percent. This compares with 25.2 percent in Germany and 21.2 percent in Britain. Work force reductions had become prohibitively costly in France, and, arguably, either productivity growth had to slow or job security had to be weakened. The latter occurred with the collective agreement of 1984, facilitating the closures and consolidations of the mid-1980s.

Trends in Capacity: Descriptive Statistics

One might expect that technological and locational cost considerations would dominate major and permanent restructuring decisions. However, as with the allocation of production, there is considerable evidence that equal proportionate reduction rules were applied at the Community level and that employment concerns were a major factor guiding investment and capacity reduction decisions within countries.

Tables 6.6 to 6.9 present descriptive statistics comparing the allocation of hot-rolling capacity in 1986 with that in 1977, which is used as the base year of comparison.[10] These summary statistics provide evidence on changes in plant size, concentration, and capacity shares. Table 6.6 shows the evolution of plant size between 1977 and 1986 by country for the plants in my sample. Because economies of scale generally are associated with fully integrated works, those with only electric steel furnaces or with only rolling facilities at the beginning of the period are excluded from the table. Smaller plant size is correlated with older equipment and suboptimal location.

The variation in plant size is considerable both within and across countries in both years. The standard deviation of capacity in hot-rolled steel is large in all countries in 1977 and, except in Belgium and Luxembourg, increases over time. The concentrations of older, smaller works in France, Luxembourg, and Britain are reflected in the relatively low mean and median plant size for those countries.[11] The competitiveness problems are most striking for Britain, which scored low in all indicators of size, despite the fact that Britain had closed a number of its least efficient plants in the mid-1970s. Its mean and median plant sizes in 1977 were by far the lowest in the Community and were on the order of a third of those in Germany. In addition, Britain's percent of steel produced in works with capacity over 1 million tons was the lowest in 1977, except for Luxembourg's.[12]

Despite a restructuring strategy of concentrating production at the most efficient works, there is little evidence of an increase in plant size at the Community level. Somewhat surprisingly, all indicators of plant size—mean, median, and the percent of steel produced in works of more than 1 and 3 million tons—remain at about the same level throughout the period. Moreover, the standard deviation of capacity increased slightly from 1977 to 1986. Thus, even with a number of closures, the dispersion in plant size was not reduced. These patterns gen-

Table 6.6. Trends in plant size between 1977 and 1986.[a]

	Mean (in thousands of tons)[b]	Median (in thousands of tons)	Percent of capacity over 1 million tons	Percent of capacity over 3 million tons
West Germany				
1977	2,443 (2,377)	1,614	93	65
1986	2,138 (2,444)	1,058	88	71
Belgium				
1977	2,603 (2,099)	1,650	100	48
1986	2,359 (950)	2,600	92	30
France				
1977	1,743 (1,515)	1,285	85	55
1986	1,939 (2,044)	1,150	90	60
Italy				
1977	3,076 (4,106)	1,640	93	67
1986	3,380 (4,462)	1,160	97	74
Luxembourg				
1977	1,199 (592)	900	61	0
1986	980 (181)	936	31	0
U.K.				
1977	887 (812)	470	70	0
1986	1,053 (947)	550	85	0
E.C. total				
1977	1,922 (2,098)	1,280	88	52
1986	1,960 (2,169)	1,120	88	55

a. All figures are based on a sample of works in the author's data base and refer to capacity in hot-rolled steel. This sample excludes plants with only electric steel furnaces and those with only rolling facilities.

b. Standard deviations are in parentheses.

Table 6.7. Evolution of capacity and concentration by product line between 1977 and 1986.[a]

	Year	Total	Plate	Hot-wide strip	Sections	Bars and wire rod	Hoop and strip
Capacity (in thousands of tons)	1977	129,669	16,817	63,614	14,368	29,046	5,824
	1986	108,329	13,117	65,856	10,432	16,908	2,016
Percent change, 1977–1986		−16.5	−22.0	3.5	−27.4	−41.8	−65.4
Number of plants	1977	87	22	21	34	61	16
	1986	71	17	19	27	38	4
Total closures among these plants, 1977–1986		16	3	3	3	10	5
Four-firm concentration ratio	1977	46.5	59.1	47.5	51.7	50.0	79.9
	1986	45.3	58.9	47.6	63.8	48.2	100.0
Eight-firm concentration ratio	1977	67.1	86.6	75.1	80.0	75.1	99.4
	1986	65.4	90.0	74.6	88.2	71.7	100.0
Herfindahl index[b]	1977	0.0751	0.1132	0.0893	0.0988	0.0891	0.1770
	1986	0.0739	0.1190	0.0891	0.1327	0.0868	0.3020

a. All calculations are based on the sample of works in the author's data base.
b. The Herfindahl index is defined as $\sum_i s_i^2$, where s_i is company i's share of capacity in the relevant product category.

Table 6.8. Comparison of shares in hot-rolling capacity, 1977 and 1986.[a]

	1977 share	1986 share
West Germany	0.372	0.378
Belgium	0.105	0.115
France	0.178	0.176
Italy	0.121	0.127
Luxembourg	0.046	0.036
Netherlands	0.048	0.049
U.K.	0.130	0.122
Correlations between 1977 and 1986 shares		
Country level (7 observations)	0.998	
Company level (29 observations)	0.986	
Regional level (37 observations)	0.949	

a. All calculations are based on the sample of works in the author's data base.

erally are mirrored within individual countries. The notable exception is Britain, which increased its average and median plant size. Although it remained below the Community average, by 1986 it had narrowed the gap somewhat.

Table 6.7 examines trends in concentration of capacity for total hot-rolled products and for individual product lines. The figures are calculated on the basis of works in my sample. Reductions in capacity over the period varied considerably across product lines. For hot-rolled products as a whole, capacity fell by 16.5 percent, but this ranged from a small increase for hot-wide strip to decreases of more than 40 percent for bars and wire rod and of more than 65 percent for hoop and strip. The large decline in bars and wire rod reflects the fact that the major producers were losing market share to the mini-mills, which are not included in the sample. Capacity reductions in hoop and strip occurred for technological reasons.

Reductions in capacity involved a number of partial and total closures. Yet despite the magnitude of the reductions, there is little evidence of an increase in concentration. The table presents four- and eight-firm concentration ratios and Herfindahl indices[13] for total hot-rolled products and for individual product lines. Although these figures overstate true concentration because they are based on an incomplete sample of firms, they nonetheless depict the overall effects of the Davignon Plan on concentration.[14] Except for the product category hoop and strip, measures of concentration are similar in the two periods. This is particularly apparent for total hot-rolling capacity. At least among the major producers, then, market shares remained fairly stable during this period of restructuring.

Table 6.8 compares the share of hot-rolled capacity for each country in 1977 with that in 1986 and presents the correlations between those shares at the national, company, and regional level for the plants in my sample. The figures display a remarkable stability of shares across time and strongly suggest that the Davignon Plan resulted in simple proportionate reductions, at least for the major producers. The correlation of total hot-rolled capacity at the national level is .998 and at the company level is .986. These correlations are high, especially given the lumpiness of the capital equipment and the large differences in competitive position across countries and companies. The large shifts in market share predicted by some analysts were not realized.

The regional level is the most appropriate level of analysis for labor markets, and for this reason the correlation at the region III level is calculated. The regional level is also an interesting one for comparison because it tends to capture locational competitiveness and the age of the works. For example, the raw materials orientation and suboptimal plant layout placed most works in regions like the Saar and Lorraine at a severe competitive disadvantage in the 1980s. However, for historical reasons these regions also tended to be the most economically dependent on steel. Social considerations mitigated capacity losses in these regions, a fact that is borne out by the high correlation of .949.

Table 6.9 breaks down country- and company-level correlations by product line. One would expect considerably more flexibility within product lines. However, with the exception of bars and wire rod and hoop and strip, the correlations remain high.

Although the official criteria in negotiations concerning capacity

Table 6.9. Correlations between 1977 and 1986 capacity shares by product line.[a]

	Plate	Hot-wide strip	Sections	Bars and wire rod	Hoop and strip
Country level	0.991	0.986	0.955	0.940	0.681
Company level	0.954	0.977	0.915	0.815	0.677

a. All calculations are based on the sample of works in the author's data base.

were financial viability in 1986 and previous government subsidization,[15] the E.C. Commission admitted that social considerations were important, albeit unofficial, criteria that required "solidarity" cuts from efficient producers. Thus, while stressing the need to allocate capacity to the most efficient works and to end all subsidization, the commission clearly understood the social pressures underlying those subsidies, and realized that for the program to be successful, these pressures had to be diffused. From the outcome, it appears that the commission resolved these conflicting forces by applying roughly a simple rule of equal proportionate sacrifice. This outcome helped preserve traditional sites of production and employment.

The Allocation of 1986 Hot-Rolled Capacity

Regression equations were estimated to formally test the determinants of the 1986 allocation of capacity. Paralleling the analysis of production allocation, the following equation was estimated:

$$(6.3) \quad q_{i86} = \beta_0 + \beta_1 q_{i77} + \sum_j \beta_2 X_{ij} + \sum_k \sum_j \beta_{3j} X_{ij} C_k + \sum_l \beta_{4l} D_{il} + \sum_k \beta_{5k} q_{i77} C_k + \varepsilon_{ij},$$

where the dependent variable is the share of the plant's hot-rolled capacity in 1986; q_{i77} is its share of capacity in 1977, taken as the base year of comparison in the capacity equations; the X_j are other exogenous variables in (6.1), including measures of the quality and vintage of the capital stock and location, interacted with q_{i77}; the C_k are country dummy variables; and the D_l are product mix variables interacted with q_{i77}.

Because a number of plants in the sample were closed by 1986, creating the possibility of biased estimates due to the truncation of the dependent variable, the reported equations were estimated using the tobit model. Columns (1), (2), and (3) of Table 6.10 report the results of equations that include plants from all countries in the study. The equations associated with columns (1)–(3) include equipment variables at the hot-rolling stage; at the hot-rolling and steel-making stages; and at the hot-rolling, steel-making, and iron-making stages, respectively. The differing number of observations across these equations is due to missing data, largely on steel and iron equipment variables, and to the fact that not all works are fully integrated.

The coefficients on the variables measuring the year of hot-rolled construction, percent of steel continuously cast, diameter of the blast furnace, and coastal location are all positive, indicating that for the E.C. overall capacity was being allocated toward plants with more modern equipment and superior location. Only the coefficients on the first two variables tend to be significant, however. The coefficients on the product mix variables are all negative, which is consistent with the omitted product line's being hot-wide strip, whose total capacity increased slightly over the period.

The results reported in columns (4), (5), and (6) include the interactions of country dummy variables with the capital equipment variables as a test for cross-country differences in how capacity was allocated. The equations reported in these columns are run on the subset of German, French, and British plants. The equation specifications parallel those reported in Table 6.4 on the allocation of production.

By the mid-1980s job security had been weakened in France, and therefore one would expect to observe greater evidence of restructuring, all else constant, in both France and Britain than in Germany. The results in Table 6.10 generally support this hypothesis. The interaction of the French dummy variable with the variable capturing the year of construction of hot-rolling equipment is positive and statistically significant in all three equations. The coefficients on the French and British country dummy interactions with the variable capturing percent of steel continuously cast are positive in both equations, and statistically significant for the U.K. term in one of the equations. The coefficient on the U.K. interaction term with the variable measuring diameter of the blast furnace is also positive, though not significant.

Together, these results suggest that the quality and vintage of the capital stock were more important criteria in determining a plant's capacity in France and Britain than in Germany, which is consistent with the weaker job security in these countries. These results are also broadly consistent with the results in Table 6.4 concerning cross-country differences in the allocation of production. The one exception is the negative and statistically significant coefficient on the interaction of the French dummy variable with the variable measuring diameter of the blast furnace.

Table 6.10. The determinants of capacity allocation.[a]

Variable	Mean[b]	(1)[c]	(2)[c]	(3)[c]	Mean[d]	(4)[c]	(5)[c]	(6)[c]
Share of hot-rolled	.011	.517	.231	.021	.010	1.041	.974	1.991[e]
capacity, 1977	(.015)	(.361)	(.330)	(.586)	(.012)	(.662)	(.683)	(.762)
Year of hot-rolled	.703	.010[f]	.009[f]	.008	.604	.004	.006	.005
construction	(.987)	(.006)	(.005)	(.007)	(.792)	(.010)	(.010)	(.010)
*France	n.a.	—	—	—	.153	.044[e]	.043[e]	.059[e]
					(.431)	(.013)	(.021)	(.013)
*U.K.	n.a.	—	—	—	.113	.002	.001	−.024
					(.278)	(.026)	(.018)	(.028)
Percent steel	.371	—	.009[e]	.009[e]	.377	—	.035	−.195
continuously cast[g]	(.694)		(.003)	(.003)	(.710)		(.369)	(.413)
*France[g]	n.a.	—	—	—	.057	—	.543	.842
					(.320)		(1.226)	(.596)
*U.K.[g]	n.a.	—	—	—	.045	—	.293	1.555[e]
					(.133)		(.526)	(.700)
Diameter of blast	.129	—	—	.028	.110	—	—	−.082[e]
furnace	(.168)			(.054)	(.141)			(.040)
*France	n.a.	—	—	—	.027	—	—	−.169[e]
					(.077)			(.069)
*U.K.	n.a.	—	—	—	.016	—	—	.365
					(.038)			(.267)
Coastal plant	.004	−.0005	.131	.130	.002	.032	−.098	.056
	(.012)	(.103)	(.103)	(.122)	(.007)	(.154)	(.274)	(.204)

Share plate	.001	−.264	−.731	−.688	.001	−.390	−.527[f]	−.543[f]
	(.003)	(.659)	(.508)	(.589)	(.003)	(.359)	(.314)	(.318)
Share sections	.001	−.434	−.537	−.383	.001	−.139	−.561[f]	−.381
	(.002)	(.347)	(.345)	(.423)	(.002)	(.298)	(.345)	(.385)
Share bars and wire rod	.003	−.679[e]	−.593[e]	−.629[f]	.003	−.827[e]	−.756[e]	−.958[e]
	(.003)	(.260)	(.292)	(.344)	(.004)	(.163)	(.177)	(.232)
Share hoop and strip	.0005	−.742	−.301	−.474	.0005	−1.684[e]	−1.692[e]	−2.583
	(.0015)	(.566)	(.411)	(.498)	(.0015)	(.613)	(.653)	(1.908)
France	n.a.	—	—	—	.003	−2.775[e]	−2.865[e]	−2.433[e]
					(.007)	(.779)	(1.183)	(1.048)
U.K.	n.a.	—	—	—	.002	−.159	−.112	−2.381
					(.004)	(1.702)	(1.214)	(2.650)

a. Dependent variable: share of hot-rolled capacity, 1986. Its mean (standard deviation) is .011 (.017) for column (1); .013 (.018) for column (2); .014 (.019) for column (3); .010 (.016) for column (4); .011 (.016) for column (5); and .012 (.017) for column 6.

b. Applies to equation results reported in columns (1)–(3). Standard deviations are in parentheses. Means and standard deviations are for the maximum number of observations. They do not vary substantially with sample size, however.

c. Standard errors are in parentheses. Number of observations is 87 for column (1); 78 for column (2); 65 for column (3); 68 for column (4); 61 for column (5); and 50 for column (6). Columns (1)–(3) report the results of equations that include plants from all countries; columns (4)–(6) report the results of equations that include only plants in Germany, France, and Britain. A constant term was included in the equations but is not reported in the tables. The plant's share of hot-rolled production in 1977 is interacted with all other right-hand variables in these equations.

d. Applies to equation results reported in columns (4)–(6). Standard deviations are in parentheses. Means and standard deviations are for the maximum number of observations. They do not vary substantially with sample size, however.

e. Coefficient significant at the .05 level or better.

f. Coefficient significant at the .10 level.

g. Coefficient and standard error multiplied by 100.

Plant Closures

A total plant closure imposes the most severe employment adjustment problems on a community. Natural attrition, early retirement, shorter working hours, and subcontracting out excess labor were important mechanisms for avoiding layoffs for plants remaining open. Closure, however, precludes these options, and necessarily entails job loss or expensive relocation. Sixteen of the plants in the sample were closed as of 1986. One would expect that companies operating in countries with relatively weak job security would be more likely to close inefficient plants, all else being constant.

Table 6.11 presents the means and standard deviations of selected variables separately for plants that remained open and for those that were closed as of 1986. The plants that closed were less competitive in key technical, location, and product mix factors. Plants that closed, on average, had older equipment in all stages of production, were smaller, and had a lower percent of steel that could be continuously cast than did plants that remained open. A lower proportion were coastal works and a higher proportion were located in regions that were highly dependent on steel for employment. In addition, in plants that were closed, on average a greater share of hot-rolled production consisted of hoop and strip and a smaller share consisted of hot-wide strip. There is wide variation in these characteristics, however, both among plants that remained open and among those that closed.

To test the determinants of plant closure, I constructed a dummy dependent variable, which equals 1 if the plant remained open and 0 if it was closed. Several dummy variables to capture the quality and vintage of the capital stock were constructed, in part to save degrees of freedom: (1) a dummy variable that equals 1 if the median year of construction of hot-rolling equipment is 1962 or earlier, indicating old equipment at the hot-rolling stage; (2) a dummy variable that equals 1 if the median year of construction of steel-making equipment is 1962 or earlier, indicating old equipment at the steel-making stage; and (3) a dummy variable that equals 1 if the median year of construction of hot-rolling and steel-making equipment is 1962 or earlier and if the median diameter of the blast furnaces is less than nine meters, indicating old equipment at all three stages of production.[16] Table 6.12 reports the results of selected equations estimated using the probit model.

Columns (1) to (3) of Table 6.12 report the results of equations that include plants from all countries in the study. According to the results

Table 6.11. Characteristics of plants that remained open, compared with those of plants that were closed.[a]

Characteristic	Mean for plants that remained open[b]	Mean for plants that were closed[b]
Ln(1977 capacity)	6.78	6.21
	(1.18)	(.96)
Year of hot-rolled construction ≤ 1962	.45	.63
	(.50)	(.49)
Year of steel-making construction ≤ 1962	.34	.56
	(.48)	(.51)
Percent of steel continuously cast	31.6	9.4
	(32.7)	(27.4)
Diameter of blast furnace (in meters)	8.1	7.2
	(2.2)	(1.8)
Equipment old at all stages[c]	.14	.29
	(.35)	(.47)
Coastal plant	.21	.13
	(.41)	(.34)
Regional dependence on steel	.73	.88
	(.45)	(.34)
Percent of hot-rolling capacity in:		
Hot-wide strip	21.6	15.9
	(38.3)	(34.4)
Plate	15.5	11.7
	(32.4)	(29.4)
Sections	15.9	16.9
	(25.1)	(36.7)
Bars and wire rod	43.0	33.8
	(40.5)	(41.7)
Hoop and strip	4.0	21.7
	(14.1)	(39.7)

a. Figures are based on the sample of plants in the author's data base.
b. Standard deviations are in parentheses.
c. A plant's equipment is classified as old at all stages if the year of construction of hot-rolling and steel-making equipment both were 1962 or earlier and if the diameter of the blast furnace was less than 9 meters.

of these equations, the vintage and quality of the capital stock appear to have an important effect on closure decisions. The coefficient on the dummy variable capturing old hot-rolling equipment is negative in columns (1) and (2) and is statistically significant in the latter, while the coefficient on the variable indicating older equipment at all three stages

Table 6.12. The determinants of plant closure.[a]

Variable	(1)	(2)	(3)	(4)	(5)	(6)
Constant	−1.052 (1.667) [−.239]	.478 (1.524) [.095]	.679 (1.475) [.134]	2.362 (2.264) [.531]	1.470 (1.938) [.272]	2.635 (2.168) [.183]
Ln(1977 capacity)	.379[b] (.197) [.086]	.038 (.231) [.008]	.077 (.214) [.015]	.050 (.233) [.011]	−.181 (.286) [−.034]	−.119 (.280) [−.008]
Share plate	.535 (.744) [.122]	—	—	−.841 (1.057) [−.189]	—	—
Share sections	.607 (.793) [.138]	—	—	−.629 (1.034) [−.141]	—	—
Share bars and wire rod	.651 (.669) [.148]	—	—	−.651 (.951) [−.146]	—	—
Share hoop and strip	−.940 (.880) [−.213]	—	—	−2.510[b] (1.210) [−.564]	—	—
Coastal plant	−.068 (.521) [−.016]	—	.555 (.710) [.109]	—	—	5.109 (437.7) [.355]
Regional dependence on steel	−.685 (.496) [−.156]	—	−.411 (.623) [−.081]	—	—	.385 (.767) [−.027]
Hot-rolling equipment ≤ 1962	−.518 (.370) [−.118]	−.868[b] (.438) [−.173]	—	−.444 (.438) [−.100]	−.762 (.511) [−.141]	—
Steel-making equipment ≤ 1962	—	−.258 (.464) [−.052]	—	—	−.008 (.606) [illegible]	—

Variable	(1)	(2)	(3)	(4)	(5)	(6)
Percent steel continuously cast[c]	—	.753 (.694) [.150]	1.156[d] (.711) [.227]	—	.583 (.749) [.108]	.876 (.752) [.061]
Diameter of blast furnace	—	.087 (.130) [.017]	—	—	.212 (.161) [.039]	—
Equipment old at all stages	—	—	−1.083[b] (.510) [−.213]	—	—	−.949 (.678) [−.066]
France	—	—	—	−1.187[b] (.594) [−.267]	−.675 (.807) [−.125]	−1.005 (.765) [−.070]
U.K.	—	—	—	−1.081[d] (.595) [−.243]	−1.187[d] (.676) [−.220]	−1.318[b] (.663) [−.092]
Number of observations	87	63	63	68	49	49
Number of plants that closed, among those observed	16	10	10	14	8	8
Log likelihood	−34.007	−23.069	−22.569	−26.395	−17.576	−16.537
Significance level	.059	.109	.075	.038	.294	.160

a. Dependent variable: 1 if plant remained open; 0 if plant closed. Standard errors are in parentheses. The numbers in brackets are the implied values of the partial derivatives of the probability that the plant remains open, evaluated at the sample mean values of the right-hand variables. Columns (1)–(3) report the results of equations that include plants from all countries; columns (4)–(6) report the results of equations that include only plants in Germany, France, and Britain. A dash appearing in the column indicates that the variable was not included in the regression equation.

b. Coefficient significant at .05 level or better.

c. Coefficient and standard error multiplied by 100.

d. Coefficient significant at .10 level.

of production is negative and significant in column (3). As expected, the coefficient on the percent of steel continuously cast is positive in both columns (2) and (3), and is moderately significant in the latter.

Plant closures were not uniformly distributed across countries. At one end of the spectrum, no Belgian plants in my sample totally closed, and only two of the twenty-seven German plants were closed. At the other end, six of the twenty-one British plants were closed, and six of the twenty French plants were closed. One explanation for the difference is that Britain and France were saddled with older, less efficient plants, necessitating more closures. If this is the case, weaker job security in these two countries would be simply a manifestation of economic necessity. An alternative explanation is that job security exerted an independent effect—acting as a constraint—on plant closures. In this case, the weaker job security in France and Britain may have enabled proportionately more closures.

As a crude test between these two hypotheses, I included country dummy variables for France and Britain in the equations used to generate columns (4) to (6) of Table 6.12. These equations include only plants located in Germany, France, and Britain. In all three equations, the coefficients on the French and British country dummy variables are negative. The French coefficient is significant in equation (4), and the British coefficient is significant in all three equations.[17]

While technical and locational factors were important determinants of closure, they probably do not explain completely the cross-country differences in closure rates. The effect on individual workers and communities inhibited plant closures in all countries, but these results suggest that social considerations carried even greater weight in countries with stronger job security.[18]

Conclusion

Substantial restructuring has occurred in the European Community steel industry over a relatively short period of time. For the major integrated producers, hot-rolled capacity fell more than 17 percent from 1980 to 1986. Capacity was reallocated from older to modern facilities and from inland to coastal works. Numerous works were partly or totally closed.

Still, employment considerations have had a profound effect on restructuring in the industry. A simple but compelling piece of evidence is the remarkable stability of production and capacity shares at the country and even the company level during restructuring. The Davignon Plan appears to have helped preserve traditional market shares, despite what many analysts believed was the scope for large shifts in these shares, were the process left to market forces. Although other factors certainly contributed to this outcome, a rule of roughly equal proportionate reductions is consistent with the strong guarantees of job security granted to steelworkers in many of the E.C. countries.

Restructuring in the E.C. steel industry may be loosely characterized as a two-step process, in which member states' traditional market shares were stabilized in the Community restructuring plan. Subject to these negotiated reductions in capacity, investment and closure decisions were made by companies with varying degrees of influence from governments. Given decentralized rationalization, one also observes the effects of job security on restructuring in cross-country differences in the patterns of production and capacity allocation and in productivity growth. The ability to concentrate production and capacity at more efficient works is, to a large degree, tied to a company's ability to reduce work force levels. The stronger the job security, the more costly are partial and total plant closures.

Evidence presented in this chapter suggests that extensive restructuring and associated productivity growth is closely associated with weaker job security. Britain provides the most direct evidence of this effect. The massive layoffs in British steel beginning in 1980 resulted in a shift of production toward more efficient facilities and in an acceleration in productivity growth despite low capacity utilization. The dramatic increase in average plant size and the large number of plant closures in Britain are further indicators of the rapid restructuring that took place during this period. Such restructuring was feasible only with relatively weak job security and large-scale work force reductions.

Restructuring in France is somewhat anomalous. Although the provisions for workers affected by reductions were fairly generous in the 1984 collective agreement compared to those in other French industries, the agreement did mark a substantial weakening of job security for workers in steel. Yet even in the 1970s, the French steel industry achieved strong productivity gains vis-à-vis Germany and Britain, although productivity growth slowed in the early 1980s. France's productivity gains are evident in the large employment reductions, discussed in the preceding chapter, and in the reallocation of production to more efficient plants, presented in the statistical analysis of Table 6.4. The employment reductions in France, which were achieved in large part through costly early-retirement measures, became prohibitively expensive as its early retirement population dwindled. One conclusion that can be drawn from the French experience is that further concentration of production at more modern works and associated productivity gains were possible only with a weakening of job security.

The weakening of job security of steelworkers in Britain and France relative to that in other countries during this period may have been partly a function of greater economic pressures to restructure, owing to older plant infrastructure and equipment and lower productivity. As the cost to governments and steel companies of maintaining production at inefficient plants, or alternatively of buying out jobs, became unacceptably high, job rights were curbed.

Although productivity and cost competitiveness no doubt played an important role in determining workers' job security, they cannot completely explain all differences in job security and restructuring strategies across countries. In tests of the allocation of production and capacity and of closure decisions for German, French, and British works, strong country differences persist, after controlling for productivity and other relevant technical and locational factors. Countries with weak job security were more likely to consolidate production in their modern, efficient facilities. This evidence suggests that worker rights to job security acted as a constraint on restructuring in the industry, and were not merely a by-product of economic pressures to restructure.

In fact, considerable scope for rationalization of production existed in most countries, including Germany. Notably, the German trade unions were leading opponents of the mergers proposed by the German government in the 1980s, because they feared these mergers would result in more plant closures and weakened job security. The extent, or at least the pace, of restructuring in the steel industry was, in part, a function of the power of labor and the job security afforded workers.

In sum, the empirical evidence presented in this book indicates that job security had profound and pervasive effects on restructuring in the European Community steel industry. Within plants, the degree of job security afforded workers affected the choice between employment reduction and hours reduction, and, quite likely, the extent of labor hoarding in the short run. More fundamentally, it affected the allocation of production and investment across plants, and consequently had a more lasting impact on the structure of production, trade, and employment in the European steel industry. A tendency to preserve traditional market shares, evident in the restructuring plan of the European Community and to varying degrees within countries, resulted in lower labor productivity and in higher production costs and capital expenditures to modernize antiquated plants.

7

Conclusion

European workers traditionally have enjoyed strong job security compared to American workers. Explicit and implicit rights to stable employment and income have been the outgrowth of paternalistic practices, collective agreements, employment security legislation, government programs, and existing social norms. Stable employment was relatively costless for employers to provide during the economic expansion and tight labor markets that characterized postwar Europe. Strong job security also has relatively little effect on resource allocation and the overall performance of the economy during an expansion. The more turbulent economic conditions of recent years, however, have highlighted the potential distributional and allocative impacts of worker rights to job security and have placed severe strains on traditional practices of job security in Europe.

The steel industry epitomizes these tensions. The steel industry is a case study of how job security affects the way in which an industry adjusts to a structural decline in demand and technical change. It is also a case study of recent trends in job security in Europe—of how its high costs to employers and the public sector have resulted in a weakening of job rights, but also of how public policies and industrial relations practices have adapted to help preserve stable employment in an unstable economic environment. Finally, it is a case study of how countries resolve trade conflicts that inevitably arise when dislocation from restructuring is exacerbated by trade.

The Equity and Efficiency of Job Security

Job security had large effects on employment, production, and investment decisions in the steel industry. In a popular sense, job security inhibited restructuring in that industry.

Did these influences result in an inefficient allocation of resources? A prerequisite for an efficient allocation of resources is that there be no obstacles to negotiating job security provisions that are efficient, given an assignment of rights. In essence, a market for job rights must be established, in which companies can buy out workers with strong job security and workers can obtain stronger job protection by accepting wage cuts or the like, if it is in both parties' interest.

Such trades occurred in contract negotiations in virtually all steel companies. Steel companies in all Continental countries offered early retirement. With a low alternative wage and worker immobility, such schemes would efficiently select out employees with fewer years remaining in their working lives. Companies in several Continental countries offered a severance payment to workers who left voluntarily. Such policies efficiently select out the more mobile, younger workers. Through early retirement and severance payments for those who leave voluntarily, companies effectively were buying out workers from their jobs. In addition, workers sometimes accepted explicit wage cuts to help finance job security measures; here workers effectively were paying companies for rights to employment. Thus, markets for job rights were feasible and widely established.

I do not argue that job security and the associated adjustment process always resulted in an efficient allocation of resources. General government assistance programs may have distorted the true price of labor. Parties to the negotiation—public and private—no doubt make mistakes, resulting in too little or too much adjustment. Unfortunately, it is impossible to isolate such effects. Rather, the point is that even in the absence of such inefficiencies, one might expect that job security would have important effects on how an industry restructures.

Distributional factors—how worker rights in jobs are defined or how the costs of adjustment are divided—are likely to influence economic decisions concerning employment, production, and investment. Individuals may attach high value to their jobs and community (apart from the income their jobs provide) and may resist change. Strong worker attachments to job and community were emphasized by all individuals interviewed for this study. Employment protection legislation and worker assistance programs, similarly, have been justified as a means of upholding European values in employment and community stability. Slower productivity growth and other possible adverse effects arising from strong worker rights to job security may be viewed, in part, as a price paid for greater employment and community stability and for a more equitable distribution of the costs and risks of economic change.

Formally, in neoclassical economic theory, when employees "consume" work and locational attributes of their job, production and consumption decisions are no longer separable, and the Coase Theorem on the independence of resource allocation to property rights assignments no longer holds. An increase in workers' rights to job security or in their share of the economic rents will be taken partly in the form of higher pay or severance payments and partly in the form of fewer layoffs and reduced hours of work. Other theories offer quite different explanations for why rights assignments may alter economic decisions. Prospect and other behavioral theories argue, in essence, that individuals' rights affect the way they perceive or "frame" a decision problem, which, in turn, influences their choices. In the case of rights to job security, workers would value their jobs more, and would experience greater loss from layoffs, the stronger their rights. Therefore, the severance payment necessary to induce workers, who have job rights, to voluntarily leave their jobs would be greater than the amount that workers, who have no job rights, would be willing to pay to keep their jobs.

The Effects of Job Security: A Review of the Evidence

Empirically, strong job guarantees exert their most direct effect on how a plant adjusts its employment levels to changes in production. In steel, plants in Continental countries, which gave workers quite strong job guarantees, adjusted employment less than those in Britain, where job security was the weakest. To a large extent, differences in the adjustment of employment levels are accounted for by differences in the adjustment of average hours per worker. Although average hours remained almost constant in Britain during the crisis, they dropped sharply after 1974 and fluctuated considerably with output thereafter in all Continental countries. This pattern of hours adjustment in Continental countries reflects the extensive use of temporary short-time work and permanent reductions in the work week or increases in vacation time.

By affecting production and investment decisions, job security also had important impacts on basic restructuring strategies. The cost of reducing labor input is likely to increase at an increasing rate with the size of the reduction. This effect is more pronounced when job security is strong and companies must rely extensively on early retirement and average hours reductions to lower labor input. Strong job security and the associated costs of implementing large work force reductions provided an economic incentive to spread necessary production and capacity cuts across plants to a greater extent than technical and locational factors alone would have warranted.

A compelling, though indirect, piece of evidence concerning the effects of job security on production and capacity allocation is the stability of national and even company shares in hot-rolled steel under the Davignon Plan. The process by which production and capacity were allocated essentially occurred in two steps. A rough rule of equal proportionate sacrifice was applied at the Community level under the Davignon Plan, and further restructuring decisions were made by companies. Certainly, there were important shifts in market shares, but these were small relative to what industry analysts believed would have been possible had the process been left entirely to market forces. Such an outcome was compatible with the strong job rights of steelworkers in many of the E.C. countries, and concerns over employment appear to have been one factor behind it.

Within countries, there is substantial evidence to suggest that worker rights to job security affected the allocation of production and capacity, including the decision to close plants. During restructuring, Britain and France significantly weakened steelworkers' job guarantees. Correspondingly, tests showed that in these countries there was a greater propensity to allocate capacity to more efficient facilities or to close plants than there was in Germany, a country with strong job security throughout restructuring. Tests also showed a clear shift in production toward more technically efficient plants within Britain in the 1980s, coinciding with the weakening of job security there. The rapid work force reductions, closures, and resulting concentration of production in more efficient plants that occurred in the 1980s are reflected in a surge of productivity in the British steel industry.

In general, strong job security resulted not only in a substitution of hours for employment adjustment, but also in the allocation of production and capacity to less efficient facilities. One result was lower productivity growth. However, there are important exceptions to these generalizations. Although job security was strong in Luxembourg, employment adjustment in Luxembourg was significantly greater than in all other countries over the 1974–1977 period; it was insignificantly different from that in Britain and was significantly greater than that in most other Continental countries over the 1974–1982 period.

The French steel industry represents another exception. Despite its strong job security through the mid-1980s, the French steel industry likewise achieved large work force reductions. Employment adjustment in France over the 1974–1982 period, though significantly less than in Britain, was greater than that in other Continental countries, except Luxembourg. In addition, analysis of cross-country differences in how production was allocated over the same period in Germany, France, and Britain suggests that the French industry was more inclined to allocate production to its technically efficient plants than was the German industry. The rapid reductions in work force levels and the allocation of production to more efficient plants show up in strong productivity growth in France over this period.

To some degree, these exceptions demonstrate that rapid restructuring and productivity improvements can be achieved without layoffs. In Luxembourg, ARBED's Anticrisis Division, in which excess steelworkers were subcontracted out to other companies or to the government, accounts for this anomaly. The large work force reductions in Luxembourg were partly an accounting phenomenon. Yet the company was fairly successful in contracting out excess workers, and its innovative program for reducing work force levels while ensuring workers jobs may have been more efficient than the work force reduction programs adopted in other Continental countries. Luxembourg's economic situation, however, was not comparable to that of other countries. Luxembourg's low unemployment made subcontracting arrangements more feasible than in most European countries, which were experiencing substantially higher unemployment.

In France, rapid work force reductions and productivity gains in the 1970s and early 1980s are indicative of the exigencies to restructure there. Certainly, rapid restructuring can be achieved despite strong job security if one is willing to pay the price. The French steel companies and government paid that price through costly job buyouts, early retirement beginning at age fifty, shorter work weeks, and internal transfers. The consolidation of capacity and the growth in productivity would have been greater, no doubt, had job security been weaker. Also significant is the fact that job rights for French steelworkers eventually were weakened. As the pool of candidates for early retirement diminished and the costs of providing job security rose, guarantees of employment in the steel industry were eliminated to enable further restructuring.

How Typical Is Steel? Job Security in Europe and Recent Trends

The strong job security enjoyed by the steelworkers is somewhat atypical in comparison with the situation in other European industries. Yet the work force reduction practices and supporting public policies are representative of the general approaches adopted in other sectors. In some instances, innovative policies in steel have been emulated elsewhere in these countries. The case of the steel industry reflects how Europeans have accommodated traditionally strong job security in a period characterized by a degree of structural adjustment and unemployment largely unknown since World War II. It also reflects how job rights have been weakened when the costs to companies and the public sector have become socially unacceptable.

The most common measures to avoid layoffs in the steel industry have been early retirement and various schemes to reduce working time, such as temporary short-time work, permanent reductions in the work week, and extended vacation time. By themselves, most Continental steel companies could not have afforded the strong job security their workers received; governments have played an integral role in financing many of these measures, and, ultimately, in guaranteeing these workers their jobs.

Early retirement and measures to reduce working time, with government backing, have been used extensively outside the steel industry to avoid layoffs or to stimulate new hiring. For example, a number of countries recently have passed legislation that encourages retirement at an earlier age. In Italy, while steelworkers are eligible for special early retirement pensions beginning at age fifty, workers affected by restructuring in other sectors may receive a pension three years prior to normal retirement (fifty-seven for men, fifty-two for women). France lowered the age at which workers receive government pensions from sixty-five to sixty in 1983. Furthermore, a system in effect in France from 1972 to 1982 guaranteed workers age fifty-five and over who had been laid off 70 percent of their wage until retirement. As a result, firms were able to reduce their work force rapidly without incurring substantial costs themselves by instituting early-retirement programs whereby retirees officially were laid off. In 1984 this provision was limited to allow older employees to work half time ("phased early retirement").

The Dutch government has passed measures allowing early retirement on a full pension at sixty-two (normally sixty-five) in certain sectors. Because unemployed workers age fifty-seven and a half and older are not required to seek employment, the steel industry and other sectors have been able to use the unemployment insurance system to help finance early retirement. Similar practices have been widespread in Germany as well, although there the government sought to curb this use of the unemployment insurance system by making companies not in economic difficulty reimburse the state. Germany and Britain also have passed special legislation to help finance early retirement or partial retirement if the company replaces the retiree with a new recruit.

In addition to financing early retirement, most countries support the use of temporary short-time work through unemployment insurance for time not worked. Moreover, in light of high unemployment, government policies in some countries have encouraged the reduction of the work week. France lowered the statutory work week from 40 to 39 hours in 1982, and the Netherlands introduced a 38-hour work week in 1985. Since 1982 the Belgian government has encouraged shorter work weeks by permitting exemptions from certain regulations on working time to companies that reduce overall working time.

For most of the postwar period, when companies' output generally was growing and labor markets were tight, the costs to companies or governments of providing stable employment were relatively low. This changed with the recessions, industrial restructuring, and high unemployment experienced by most European countries since the mid-1970s. As the costs of providing job security soared (and as labor's bargaining power weakened in a period of slack demand), job rights have been weakened in limited though important ways. In the steel industry, this occurred most notably in Britain and France.

The weakening of job rights is more generally evident in many European countries with the loosening of legal restrictions on collective and individual dismissals. Although changes in employment protection laws have not fundamentally altered job rights, they have lowered the costs to companies of laying off workers and have introduced some important exemptions concerning the workers covered by these statutes. For example, the Belgian government significantly reduced the notice period required for white-collar workers,[1] and allowed plants in economic difficulty to spread mandatory severance payments over time rather than provide them in a single lump sum. In the Netherlands, the administrative procedures for laying off workers have been simplified; and in France, government approval of layoffs has been eliminated altogether. Germany has raised the percent of the work force that must be laid off before a firm must comply with collective dismissal laws and has provided some relief for new and small firms.

Restrictions on the use of fixed-term contracts and temporary workers typically accompany legal restrictions on dismissals. Some of the most important legal changes have involved the relaxation of restrictions on the use of temporary workers and fixed-term contracts, and thereby the coverage of dismissal law. In Germany, France, and the Netherlands, changes in labor laws allow firms greater flexibility to hire and fire workers by increasing the permissible length of fixed-term contracts or by expanding the permissible use of temporary workers.

One response to the high cost to employers and the public sector of providing job security, therefore, has been to weaken job security itself. Another response, however, has been the adoption of various measures designed to lower labor costs and thus compensate for the high cost of job security. These measures have been the result of both government and private initiative. Obviously, wage concessions may be granted in return for job security, and such trade-offs were common in the steel industry. In addition, a quid pro quo for reduced working time to preserve jobs has often been greater flexibility as to when those hours are scheduled and how they are paid. In Luxembourg, workers negotiated nominal wage reductions, which were partly compensated through additional vacation time. Tied to increased time off was a provision allowing ARBED to bank these vacation days from year to year. ARBED may stock up to twenty-two days per worker over a three-year period. Such flexibility allows the company to avoid paying overtime or avoid making new hires when conditions improve.

In Germany, although the reduction in the work week in the steel and metalworking industries was negotiated at the sectoral level, details were decided at the firm level. A survey of employers after the 1984 reduction showed that most employers used a variety of measures to reduce working time in a flexible way: 75 percent gave an extra day of leave; 17 percent varied the reduction in working time according to the category of worker involved; and 50 percent averaged the work week over a two-month period.[2]

Outside the steel industry, measures to relax restrictions on night and weekend work,[3] to calculate overtime on a long-term basis, and to otherwise increase the flexibility of the work force have gained popularity in Europe as part of a broader attack on low productivity growth, high labor costs, and high unemployment. For example, in Belgium, the Hansenne experiment, introduced in 1982, permitted exceptions to a number of regulations governing overtime and night work, provided that companies also reduced overall working time and created jobs. The success of the Hansenne experiment led to the passage of a law in 1985 allowing the calculation of working time on a longer-term basis, subject to union concurrence. In a national agreement in the Netherlands, unions and employers agreed to a freeze on real wages and a reduction in working time, but also to a reorganization of working time through such measures as weekend work and flexibility in calculating working time. In France, the law that lowered the statutory work week to thirty-nine hours also allowed the averaging of hours across weeks, greater flexibility in the use of overtime, and exceptions to Sunday work restrictions. And in Germany, the movement toward shorter work weeks coupled with greater flexibility in the scheduling of hours, which began in the metalworking sector, has quickly spread to other industries.

In sum, the crisis of the E.C. steel industry mirrors the deeper troubles experienced by European economies in the 1970s and 1980s. The methods of work force reduction in steel, in turn, reflect private and public sector approaches to handling labor adjustment in the member states during this volatile economic period. Historically, workers in Europe have enjoyed relatively stable employment and strong job rights. The evolution of work force reduction practices and related public policies has been a natural outgrowth of a tradition of strong job security. As the costs to employers of providing job security have risen, they have been shifted, to a large extent, onto the public sector through such measures as subsidies for early retirement and unemployment insurance for short-time work. As the costs to companies and governments of strong job security have grown unacceptably high, job security has diminished through a weakening of provisions in collective agreements and the relaxation of employment protection laws.

But paralleling a decline in employment security has been the growth of a wide variety of measures to increase the internal flexibility of the work force. Such measures help compensate employers for the costs of providing job security, and help compensate, to some extent, for the adverse consequences that strong job security may have on an industry's ability to restructure. As such, these measures represent an adaptation to the industrial relations system to help preserve employment stability for workers in a period of greater economic turbulence and uncertainty.

Restructuring in the E.C. steel industry also may be broadly representative of solutions to trade conflicts in open economies. As traditional trade barriers within the European Community and between the Community and its trading partners are eliminated, the resolutions of trade conflicts increasingly are in the form of managed trade agreements, like the Davignon Plan. Such agreements, almost by definition, tend to preserve the status quo. Although the Davignon Plan achieved its ultimate goal of substantially lower capacity, it did so by applying, roughly, a rule of equal proportionate sacrifice. By implication, this has resulted in costly production and investment in antiquated and suboptimally located plants. With strong worker rights in jobs, coupled with low labor mobility and high regional unemployment, such a solution may have approximated an efficient one. However, it has had a profound, possibly lasting effect on the patterns of production, trade, and employment in the steel industry.

Appendix

Notes

References

Index

Table A.2. France

Early retirement	Transfers	Short-time or reduction in work week	Incentives for voluntary departure or benefits to laid-off workers
1984 agreement	*1984 agreement*	*Short-time*	*1979 agreement*
Age 50. Mandatory.	Transfers could take place across companies. Workers had right to one refusal. Some protection of seniority and income for both internal and external transfers. Workers older than 45 guaranteed employment in same region, if possible.	State provided compensation for hours not worked.	50,000 francs for voluntary departure.
Benefits		*Reduction in work week*	*1984 agreement: reconversion program*
Ages 50–55: 75% of gross wage. Ages 55–60: 70% of gross wage. Older than 60: normal early retirement benefits.		As of 1983: 37.5-hour work week in steel overall; 35-hour week in some works.	Two-year program, during which worker received 70% of former gross salary and was still formally employed by steel company. If after 2 years worker did not find permanent employment, company was obligated to ensure that he receive 2 offers, with intention that one be in same region. Agreement established a program to reindustrialize steel regions.
Funding			*Mobility allowance*
State: Covered about 90% of costs of program. *E.C.:* Reimbursed state for some of costs.			Provided by state and E.C.

Table A.3. The Netherlands

Early retirement	Transfers	Short-time or reduction in work week
Special three-time program at age 57.5.	*Income guarantees*	*Short-time*
Funding	Worker kept former salary and salary scale. Shift allowance gradually reduced until age 57.5. *E.C.* helped defray the cost.	Funding supplied by state, with E.C. contribution under special social funds. Company supplemented state benefits.
State: Paid 70% of gross wage from unemployment insurance fund.		*Reduction in work week*
Company: Supplemented state benefits to guarantee 87.5% of previous net pay and contributed to pension, to ensure full vesting upon normal early retirement.		In 1986 average work week for shift workers lowered to 36 hours; other employees worked 40 hours per week, but received 13 additional holidays.
E.C.: Maximum of 5,000 ECUs per worker for pension contribution.		

Table A.4. Belgium

Early retirement	Transfers	Short-time or reduction in work week	Incentives for voluntary departure or benefits for laid-off workers
Age 53–55. Voluntary. *Benefits* Workers received state unemployment insurance, plus supplementary payments from company. Workers typically received 70–80% of previous net wage. *Funding* *State:* Unemployment insurance system. *E.C.:* 50% reimbursement to state. *Company:* Supplementary payment, but could take out 15-year loan for these payments from state. State pays interest during first 5 years; in sixth year, pays 5% of interest.	*Income maintenance* Period varied by company. *Consultation* Generally, worker was allowed one refusal. Could be asked to move to a different site but usually not to a different region. *Funding* *State contribution to income maintenance:* Topping-up allowance, to bring worker up to 100% of reference wage. (Reference wage was 72,100 Belgian francs per month in 1982.) *E.C.:* Reimbursed state for 50% of costs.	*Short-time benefits and funding* *State:* 1,000 Belgian francs per day. *Company supplement:* About 175 Belgian francs per day. For hours not worked, average reduction was 18–20% of net wage. *Reduction in work week* Work week reduced from 40 hours to 35–37 hours.	*Income maintenance* Supplementary unemployment insurance benefits in the event of unemployment. Topping-up allowance to bring worker up to reference wage for 14 months in event of reemployment. Costs shared by state and E.C. *Mobility allowance* Provided by state and E.C.

Special plan at Cockerill-Sambre, 1984–1985
Age lowered to 53 from 55. Financed in part by 2% wage cut in 1984 and 1985.
Social security scheme
For workers not classified as redundant. Age 58 for manual workers; 55 for handicapped workers. Benefits for these workers somewhat lower than for those classified as redundant.

Table A.5. Italy

Early retirement	Transfers	Short-time
Age 50 (lowered from 55 in 1984). Voluntary.	*Internal*	*Benefits*
Funding	Workers guaranteed former earnings for 18 months.	80% gross income for hours not worked.
Company: Tax on gross salary of all employees paid into early retirement fund.	*Reemployment in public works*	*Funding*
State: Supplemented early-retirement fund with transfer from CIG.	Workers received salary about 20% higher than public sector salary for 18 months.	*State:* CIG, which is partly financed from tax on establishments with 50 or more employees.
E.C.: Contributions from special E.C. social funds.	*Funding*	*E.C.:* Contributions from special E.C. social funds.
	Costs shared by state and E.C.	

Table A.6. Luxembourg

Early retirement	Transfers	Short-time or reduction in work week	Incentives for voluntary departure
Age 57. Compulsory from 1978 to 1980. Voluntary after 1980. Special one-time early retirement for handicapped workers age 50 and older. *Benefits* 85% of former wage for the first year. 80% of former wage for the second year. 75% of former wage for the third year. Normal early retirement benefits thereafter. *Funding* Government passed law to fund early retirement with unemployment insurance fund. Financed by "solidarity tax" on income. E.C. contributed about 50% of cost above state unemployment benefit.	1975–1976: ARBED put extra workers in government program "Work for General Interest." 1977: Anticrisis Division (DAC) formed. Extra workers contracted out to other firms and to the government. Workers guaranteed their former wage. *Funding* E.C. and state covered most of company's losses on DAC.	Increased vacation time in lieu of reduction in work week.	Severance payment for any worker finding a job in another industry: about 1 year's pay. *Guaranteed income support in event of reemployment* 95% of former wage for the first 6 months. 90% of former wage for the next 6 months. 85% of former wage for the next 6 months. *Funding:* Supplied by state, with 50% refund from E.C.

Table A.7. Britain

Early retirement	Transfers	Short-time or reduction in the work week	Incentives for voluntary departure	Benefits for laid-off workers	Retraining
Age 55 for men; 50 for women. Men 55 or older and women 50 or older who were laid off received nonactuarily reduced pension. E.C. helped defray the costs of early retirement.	*Income maintenance* Workers guaranteed 90% of previous gross earnings for 18 months if under age 55; for 24 months if 55–60; for 30 months if older than 60. E.C. helped defray the costs of income maintenance.	*Short-time* State funding of short-time stopped in 1984. By collective agreement, company guaranteed workers on short-time 80% of normal shift income. *Reduction in work week* Work week lowered from 40 to 39 hours in 1983.	*Voluntary* Severance payments for voluntary departure occasionally used. *Mobility allowance* Costs shared equally by state and E.C.	*Severance payment* Basic rate in steel was 150% of minimum statutory rate. Additional payments negotiated at plant level. *Income maintenance* Company supplemented the unemployment insurance benefit. Amount and period varied by age. E.C. reimbursed company for some of these expenses. *Pension option* Men 55 or older and women 50 or older could invest company unemployment insurance supplemental benefits in their pension. *Mobility allowance* Provided by state and E.C.	Fifty-two weeks of retraining at 100% of former gross earnings, plus pension contributions from steel company. *Funding* Supplied by state, with 50% reimbursement from E.C.

Notes

1. Introduction

1. See, for example, Flaim and Sehgal (1985); and Podgursky and Swaim (1987a, 1987b, 1987c).

2. This was possible in all countries except Italy.

3. See, for example, Diebold (1959); Commission of the European Communities (1987b); Harris (1984); Sirs (1980); and Flaim and Sehgal (1985).

2. The Response to Excess Steel Capacity in the European Community

1. Spain and Portugal subsequently joined the Community, but their steel industries were not covered by common market policies during restructuring under the Davignon Plan.

2. Presumably a situation of excess capacity would result from insufficient oligopolistic collusion or state intervention. It was feared, however, that in the event of such a crisis, producers would reestablish the powerful prewar cartels.

3. Imports traditionally accounted for a small percentage (about 5 percent) of E.C. consumption. Surges in imports have occurred at various times since 1974, however, and have had important effects on prices.

4. In the late 1980s, the appreciation of European currencies with respect to the U.S. dollar further undermined their ability to compete both with U.S. producers and with Third World countries, whose currencies did not similarly appreciate.

5. For example, in the early years of the crisis, analysis of the British steel market focused on whether the British Steel Corporation would be able to meet British demand when the upswing occurred or whether it would, once again, lose market share because of delays in restarting furnaces.

6. According to one E.C. official, however, such a complaint was never filed.

7. Whereas the Treaty of Paris prohibits state aid, the Treaty of Rome allows the E.C. to grant exceptions. The commission used this legal loophole to formally recognize and limit state subsidies.

8. In 1981 the E.C. Commission estimated excess capacity at 50 million tons.

9. Recently, with the improvement in the steel market in Europe, there has been some increase in cross-national investments.

10. Subsequently, the French company Usinor-Sacilor acquired controlling interest in steel production in the Saar region.

3. Job Security and the Methods of Work Force Reduction

1. For example, one German unionist recalled a time when a firm had blocked construction of a railroad bridge over a river, because the train would have allowed its workers to commute to the next town.

2. One German trade unionist insisted that for many years the companies had been telling their workers that they had good jobs, that workers should stay in the same community, and that they should send their children to the companies' mills. With today's decline, the companies want the workers to move, but the workers refuse.

3. On the development of British law, see Hepple (1986). Meyers (1964) also contains interesting historical background on the legal treatment of dismissals in Britain, France, the United States, and Mexico.

4. Also a laid-off worker may bring a suit under Italy's laws governing individual dismissals.

5. The ECU (European Currency Unit) is a basket of the member states' currencies. At 1985 exchange rates, this aid is equivalent to about $108.2 million.

6. In May 1990 IG Metall negotiated a 35-hour work week to take effect in the metal-working industry, which includes steel, in 1995.

7. Between the ages of fifty and fifty-five, workers fell into a special category whereby they were still formally employed but did not work.

8. Here I refer only to the companies covered by the study. In particular, the study excludes mini-mills, which are an important component of the Italian steel industry.

9. This study covers only the nationalized steel industry—the British Steel Corporation.

10. According to internal company documents, however, such "voluntary redundancy" for the purposes of early retirement was common among white-collar workers.

11. Published data are not available for the Netherlands.

4. Why Job Rights Affect Resource Allocation: Competing Theories

1. Differences in job security across countries may also reflect differences in the total costs to be borne, arising from, say, differences in the competitive position of the steel industry across countries. In the empirical analysis in subsequent chapters, I take this possibility into account.

2. Workers may be regarded as collecting positive rents from the job. To the extent that valuation of certain job attributes grows over time, so do the rents.

3. For example, workers in declining industries often incur capital losses on their homes when they move. If they were collecting positive rents in consuming these housing services, then the capital loss underestimates the true loss.

4. The second constraint would not be binding with a drop in product price, and the problem may be written as follows:

$$(1) \quad \max_{w_1, s_1, L}. \quad PQ(L) - w_1 L - s_1(\bar{L} - L) + \lambda L[v(w_1) - v(\bar{w})] + \mu(\bar{L} - L)[v(s_1 + w_2) - B - v(\bar{w})],$$

where λ and μ are the Lagrange multipliers on the utility constraints. The first-order conditions for profit maximization are

$$(2) \quad P\left(\frac{\delta Q}{\delta L}\right) = w_1 - s_1$$

$$(3) \quad \frac{\delta v}{\delta w_1} = \frac{1}{\lambda}$$

$$(4) \quad \frac{\delta v}{\delta s_1} = \frac{1}{\mu}$$

$$(5) \quad v(w_1) = v(\bar{w}), \text{ or } w_1 = \bar{w}$$

$$(6) \quad v(s_1 + w_2) - B = v(\bar{w}).$$

Combining (2) and (5) yields

$$(7) \quad P\left(\frac{\delta Q}{\delta L}\right) = \bar{w} - s_1.$$

5. Comparative static results show that when workers have job rights, the increase in severance pay, s_1, that a firm would have to make to workers that it lays off for an incremental increase in nonmonetary costs of layoff, B, is

$$\frac{\delta s_1}{\delta B} = \left(\frac{\delta v(w_2 + s_1)}{\delta s}\right)^{-1},$$

whereas, when workers do not have job rights, the pay cut they would be willing to accept to keep their job, s_2, for an incremental increase in nomonetary costs of layoff is

$$\frac{\delta s_2}{\delta B} = \left(\frac{\delta v(\overline{w} - s_2)}{\delta s}\right)^{-1}$$

Note that $w_2 + s_1 > \overline{w} > \overline{w} - s_2$. That is, when workers have strong job rights, their monetary income is greater than it is when they do not have rights. With utility functions that are concave with respect to income, this implies that

$$\frac{\delta s_1}{\delta B} > \frac{\delta s_2}{\delta B}.$$

Therefore, for any positive B, $s_1 > s_2$, and adjustment of employment levels is less when workers have rights to their jobs than when they do not.

6. In a general equilibrium framework, property rights affect income distribution, which in turn influences resource allocation through its effects on demand and prices.

7. I do not allow average worker hours to vary in the model. In a similar model in Houseman (1988), I study how the division of rents between labor and capital affects the adjustment of employment levels and average hours. This model produces results that are qualitatively similar to those presented below.

8. The marginal rate of substitution between capital and labor and the elasticity of labor with respect to output are likely to depend not only on the effective quantity of capital, but also on the technology embodied in that capital.

9. See Azariadis (1979) and Azariadis and Stiglitz (1983) for summary discussions of these models. Although the model in this chapter does not treat uncertainty, it could be modified to do so. The introduction of uncertainty into the model would not affect the qualitative nature of the results presented below.

10. In the property rights model, I assumed that all individuals were treated equally. Consequently, by construction, there was no involuntary unemployment.

11. In Houseman (1988), which incorporates the adjustment of average hours, I show that an increase in $\bar{V}$ or B will result in greater adjustment of average hours per worker and less adjustment of employment levels.

12. I also have shown that when there is international trade in an industry that is declining, a government may optimally subsidize its industry. In the presence of adjustment costs, government subsidies increase production and employment in the home country and shift some of the burden of adjustment onto foreign countries. See Houseman (1985).

13. See Kahneman and Tversky (1979) and Tversky and Kahneman (1986).

14. MacLean and Mills (1988) argue that prospect theory is the most con-

vincing explanation of the observed impact of property rights on decisions, in part because of its ability to explain a variety of phenomena.

15. For a discussion of the implications of these various economic phenomena for the Coase Theorem, see Kelman (1979).

16. It is assumed that state-contingent contracts do not fully solve any problems of *ex post* immobility of labor and capital. Given the fact that the steel crisis was largely unanticipated, state-contingent contracts are considered particularly unlikely in this case.

17. The models developed in this chapter are not precisely equivalent: in the property rights model an individual is guaranteed a particular level of utility with certainty, whereas in the contract model that level of utility is only an expected level. Were this difference in modeling to be eliminated, the two formulations would be equivalent.

5. Job Security and the Adjustment of Employment in Steel

1. These represent capacity levels negotiated in 1984 that were to be achieved by the close of 1985. In some cases they may differ from actual 1986 capacity levels.

2. Although there is a consensus on the existence of large economies of scale in the steel industry, this opinion is not without its dissenters. Scale economies can be reaped only with high capacity utilization, which is a function of the level of orders and the coordination of product flows. Product-demand fluctuations and limited management capabilities may render technical economies of scale economically inefficient. Pol Boel, president of Usines de Gustave Boel, a moderate-sized integrated Belgian works, is an outspoken proponent of the view, in his words, that "small is beautiful." Boel uses his surprisingly good profit performance during the crisis as testimony to his motto.

3. A problem in constructing any summary measure is the fact that capacity utilization was quite low during this period. A number of works had very old equipment that, most likely, was rarely used. I suspect that the median measure used in the empirical analysis provides a better picture of the vintage of the capital stock utilized than would a measure based on the means of capacity.

4. These estimates are described in Commission of the European Communities (1984).

5. Furthermore, the series on regional wages is incomplete over the relevant period, and wages for the years 1974 and 1982 were estimated. Estimates were made using regional wages of the proximate year and applying the average national change in wages.

6. List prices by product category for each E.C. country are published in *Iron and Steel Yearbook*. An annual average product price was developed for each works by taking a weighted average of these published prices, where the weights were the share of production in each product category for that particular year for that works.

7. This conclusion could also result from a principal-agent problem: ownership is separate from management, and information on managerial performance is costly for stockholders to obtain. As a result, managers may maximize their own utility rather than company profits.

8. Equations also were estimated with the interaction of the change in the log of production and the country dummy variables. These equations produced results that are qualitatively similar to those reported below. Although it may be desirable to include both country dummy variables and their interactions with the change in the log of production, the small number of observations precludes this specification.

9. In regressions with iron, steel, and hot-rolled production entered separately, the coefficients on these variables were not significantly different from one another.

10. The change in the log of output was instrumented with the plant's share in various product lines in 1974, with dummy variables indicating whether the plant had a coastal location and whether the region was highly dependent on steel for employment, with country dummy variables, with a measure of the plant's labor productivity in 1974, with various measures of the quality and vintage of the capital stock, and with a dummy variable indicating whether production increased or fell during the period. The inclusion of the last variable is justified on the grounds that, during a period in which the overwhelming majority of plants were experiencing substantial declines in production, the growth dummy variable is capturing an unmeasured quality of the capital stock or plant management. Its exclusion as an instrument, however, does not substantially alter the results presented.

11. The reason for their exclusion is that closure constrains the labor-output elasticity to equal 1.

12. It is, descriptively, fairly accurate to write in terms of work force reductions, particularly over the 1974–1982 period. Even for most plants where production actually increased, employment fell over the period, reflecting productivity gains. Employment actually increased from 1974 to 1982 for only two of the plants in the sample, both in Italy.

13. In Tables 5.3 and 5.5 the difference between the coefficient on the Luxembourg term and those on the German, Belgian, and Italian terms is statistically significant in some of the equations.

14. These data were obtained from the metalworkers confederation Force Ouvrière. Their overall totals are consistent with other published data.

15. Other measures of the quality and vintage of the capital stock, including the percent of capacity accounted for by open-hearth furnaces, were included in regression equations not reported here. These measures had no significant effect on employment, and their exclusion from the regression equation does not affect the results reported.

16. Hours data for the Netherlands have not been published.

17. The very sharp drop in average hours in Italy in 1983 is probably due to

the fact that some of the workers classified on short-time were working no hours. Employers use this feature of the CIG, described in Chapter 3, in lieu of layoffs.

18. The Netherlands is excluded because hours data for that country have not been published, and Italy is excluded because production for the country as a whole increased over the period.

19. The study reported by Harris (1984) finds that many displaced steelworkers were reemployed by these subcontractors at considerably lower wages.

20. Unfortunately, accurate data on the degree of subcontracting are difficult to obtain. In Chapter 6, I present some international productivity comparisons on the steel industry developed by the U.S. Bureau of Labor Statistics. However, the bureau terminated this series in 1982, in part because of the lack of reliable data on subcontracting.

21. The job rights were not precisely the same: Luxembourg steel workers lacked an occupational job right, because workers in the Anticrisis Division were employed in nonsteel activities. The point is not moot in certain instances, as is illustrated by an anecdote from Germany. When asked why German companies did not adopt a system similar to Luxembourg's Anticrisis Division, a representative of IG Metall responded that the arrangement bore too close a resemblance to forced labor in Nazi Germany.

6. Job Security and Restructuring: The Allocation of Production and Capacity in Steel

1. The transportation cost factor for raw materials could be incorporated into the model in a more sophisticated way. By excluding raw materials input from the production function, I implicitly am assuming that only capital and labor are substitutes in the production process. However, technological improvements to capital equipment may save not only on labor input, but also on raw materials. If this is the case, then plants with higher transportation costs may have a greater incentive to invest in capital technology that saves labor and raw materials, with implications for the employment adjustment model estimated in Chapter 5. The importance of a plant's location or transportation costs for employment adjustment was tested; however, no significant effect was found.

2. Missing values, particularly for Britain, probably overstate average size of the diameter of the blast furnace.

3. A plant's production share in 1974 is interacted with all of the other right-hand variables. This functional form implies that the effect of a change in an independent variable is proportional to a plant's initial share of production. Other functional forms were tried, and this one performed the best.

4. Theory suggests that regional labor market conditions, which capture the opportunity cost of labor, also would affect allocation decisions. However,

measures of regional labor market conditions were not statistically significant in any of the equations estimated and are not included in the equations reported in Table 6.4.

5. Equations also were estimated that allowed for such shifts for all countries in the 1980s. These additional variables were not statistically significant and did not affect the results on the U.K. interaction terms.

6. I also estimated similar equations for each year of data, yielding eight regressions for each year from 1975 to 1982. The patterns of the coefficients across years showed the same trends as those obtained from pooling the data, though the results from the cross-section equations tended to be less significant, probably owing to the smaller number of observations.

7. Equations that include all plants in the sample yield similar results, except that the coefficients on certain interactions for countries with relatively few observations were unstable across equations.

8. The coefficient on the initial share rises across equations, suggesting either that fully integrated plants were allocated greater shares or that missing values were correlated with older plants.

9. This can be seen by summing the coefficients on the two U.K. interaction terms for each of the equipment variables. The equations do not include country interactions with the coastal location variable. The reason for this omission is that Germany has only one coastal works.

10. The year 1977 is used as the base year of comparison for capacity because, among other things, it is the earliest date for which measures of the quality and vintage of the capital stock are available. However, as discussed in the previous chapter, some lag from the actual downturn is justifiable on the grounds that most changes in capacity between 1974 and 1977 would have been based on decisions made prior to the crisis.

11. The Belgian industry also was plagued by small, inefficient works. This is not revealed in the statistics, however, because a number of works were consolidated under a single management with the reorganization of the industry in the mid-1970s, and, somewhat artificially, groups of works were classified as a single operation.

12. The figure of 1 million tons is somewhat arbitrary and the low percentage of capacity above 1 million tons in Luxembourg is a result of the fact that a number of works fall just below that threshold.

13. The Herfindahl index is defined as $\sum_i s_i^2$, where s_i is company i's share of capacity.

14. I do not have good data on capacity for other plants in 1986. Through 1982, the plants in my sample consistently accounted for about 80 percent of production and capacity in the Community.

15. Companies receiving large government subsidies were to be penalized in terms of final capacity allocation. Such a penalty makes sense, *ex post*, because it discourages future noncooperation. Although the E.C. Commission

maintained that subsidization was a criterion, the descriptive statistics suggest otherwise.

16. The year 1962 represents an age of fifteen years from the base year, 1977. Although this cutoff is somewhat arbitrary, it was used to classify equipment as old in Montenero (1982).

17. The reason for the drop in significance on the French dummy variable may be due to the fact that a number of observations in France are lost when steel and iron equipment variables are included.

18. This conclusion finds support in common criticism of German economic policy, which argues that it has devoted too many resources to propping up declining industries in the northern regions, which are heavily dependent on sectors like steel and mining, rather than forcing these regions to restructure their economic bases.

7. Conclusion

1. Often, dismissal laws in Europe distinguish between white- and blue-collar workers, providing greater protection to the former.

2. Two surveys of employers were conducted, one by the German employers' association and the other by the union. The results reported in the text come from the employers' survey, though the union survey produced similar findings. For further discussion of these surveys and their results, see Rojot (1989).

3. Owing to the fact that it is a continuous process industry, the steel industry does not generally have restrictions on night and weekend work.

References

Adams, Walter. 1982. "The Steel Industry." In Walter Adams, ed., *The Structure of American Industry,* 6th ed. New York: Macmillan.

Azariadis, Costas. 1979. "Implicit Contracts and Related Topics: A Survey." In Z. Hornstein, et al., eds., *The Economics of the Labor Market*. Proceedings of a conference on the labor market, sponsored by Her Majesty's Treasury, Department of Employment and the Manpower Services Commission at Magdalen College, Oxford. September 10–12.

——— and Joseph E. Stiglitz. 1983. "Implicit Contracts and Fixed Price Equilibria." *Quarterly Journal of Economics* 98 (Supplement): 1–22.

Blanpain, Roger, ed. 1980. *Bulletin of Comparative Labour Relations: Job Security and Industrial Relations*. Bulletin 11. Boston: Kluwer Law and Taxation Publishers.

Coase, Ronald. 1960. "The Problem of Social Cost." *Journal of Law and Economics* 3 (October): 1–44.

Cockerill, Anthony. 1985. "Downward Adjustment in the Steel Industry: The Case of the UK." Mimeo. August.

Commission of the European Communities. 1978. *ECSC Redevelopment Policies*. SEC(78) 4351 (final). Brussels.

——— 1983. *General Objectives Steel 1985*. COM(83) 239 (final). Brussels. April 22.

——— 1984. *The Regions of Europe*. COM(84) 40 (final/2). Brusssels. April 4.

——— 1985a. "Stronger Community Structural Measures to Assist Steel Restructuring Areas." COM(85) 384 (final). Brussels. July 17.

——— 1985b. "Rules on Aid and Financial Transfers to the Steel Industry after 1985." COM(85) 376 (final). Brussels. July 19.

——— 1985c. *General Objectives Steel 1990*. COM(85) 450 (final). Brussels. July 31.

——— 1985d. "Introduction of Production Quotas under Article 58 of the ECSC Treaty after 31 December 1985." COM(85) 509 (final). Brussels. September 25.

——— 1986. "Intermediate Review 1986: On the Realization of the General Objectives Steel." Brussels. June 10.

——— 1987a. "Current Position and Prospects for the Community Steel Industry." Original in French. Brussels. March 3.

——— 1987b. "Commission Communication to the Council: Steel Policy." COM(87) 388 (final/2). Brussels. September 17.

——— 1987c. "Communication from the Commission to the Council Amending COM(87) 388 final/2 of 17 September 1987: Steel Policy." COM(87) 640 (final). Brussels. November 26.

Crandall, Robert W. 1981. *The U.S. Steel Industry in Recurrent Crisis: Policy Options in a Competitive World*. Washington: Brookings Institution.

Diebold, William. 1959. *The Schuman Plan: A Study in Economic Cooperation, 1950–59*. New York: Praeger.

EUROSTAT. *See* Statistical Office of the European Communities.

Flaim, Paul O., and Ellen Seghal. 1985. "Displaced Workers of 1979–1983: How Have They Fared?" *Monthly Labor Review* 108 (June): 3–16.

Groupement de la Sidérurgie. 1985. "Measures Sociales d'Accompagnement pour la Restructuration de l'Industrie Sidérurgique (CECA)." Mimeo. Brussels. September 30.

Harris, C. C. 1984. "Fate of Redundant Steelworkers." Presented at a seminar at Port Talbot, Wales, sponsored by University College Swansea, Wales, Department of Adult and Continuing Education. November 24.

Hepple, Bob. 1986. "Great Britain." In Roger Blanpain, ed., *International Encyclopaedia for Labour Law and Industrial Relations*. Supplement 63 (March). Boston: Kluwer Law and Taxation Publishers.

Hogan, William T. 1983. *World Steel in the 1980s: A Case of Survival*. Lexington, Mass.: Lexington Books.

Houseman, Susan N. 1985. "Job Security and Industrial Restructuring in the European Community Steel Industry." Dissertation, Harvard University.

——— 1988. "Shorter Working Time and Job Security." In Robert A. Hart, ed., *Employment, Unemployment, and Labor Utilization*. Boston: Unwin Hyman.

——— 1990. "The Equity and Efficiency of Job Security: Contrasting Perspectives on Collective Dismissal Laws in Europe." In Katharine Abraham and Robert McKersie, eds., *New Developments in the Labor Market*. Cambridge, Mass.: MIT Press.

Kahneman, Daniel, and Amos Tversky. 1979. "Prospect Theory: An Analysis of Decision under Risk." *Econometrica* 47 (March): 263–291.

Kelman, Mark. 1979. "Consumption Theory, Production Theory, and Ideology in the Coase Theorem." *Southern California Law Review* 52 (March): 669–698.

Knetsch, Jack L., and J. A. Sinden. 1984. "Willingness to Pay and Compensation Demanded." *Quarterly Journal of Economics* 99 (August): 507–521.

Leibenstein, Harvey. 1966. "Allocative Efficiency vs. X-Efficiency." *American Economic Review* 56 (June): 392–415.

Leontief, Wassily. 1946. "The Pure Theory of the Guaranteed Annual Wage Contract." *Journal of Political Economy* 54 (February): 76–79.

MacLean, Douglas, and Claudia Mills. 1988. "Norms and Behavior in Philosophy and Social Sciences." Mimeo. Center for Philosophy and Public Policy, University of Maryland, College Park.

Maury, Jean-Michel. 1985. "Labour Market Flexibility in the Member States of the Community." Original in French. Paris: Center for Research and Information on Technology, Economics, and the Regions. September.

Meyers, Frederick. 1964. *Ownership of Jobs: A Comparative Study.* Los Angeles: Institute of Industrial Relations, University of California.

Montenero, A. 1982. "Les Installations Sidérugiques de la C.E.E. au 01.01.1981." Brussels: Commission of the European Communities. Enquête rapide des services de la commission. July.

Organization for Economic Cooperation and Development. 1984. "Manpower Policies to Facilitate Structural Change: Report by a Consultant." Mimeo. Steel committee.

Podgursky, Michael, and Paul Swaim. 1987a. "Health Insurance Loss: The Case of the Displaced Worker." *Monthly Labor Review* 110 (April): 30–33.

——— 1987b. "Job Displacement and Earnings Loss: Evidence from the Displaced Worker Survey." *Industrial and Labor Relations Review* 41 (October): 17–29.

——— 1987c. "Duration of Joblessness Following Displacement." *Industrial Relations* 26 (Fall): 213–226.

Rojot, Jacques. 1989. "National Experiences in Labour Market Flexibility." In *Labour Market Flexibility: Trends in Enterprises*. Paris: Organization for Economic Cooperation and Development.

Sirs, William. 1980. "The Problems of Labour Force and Readaptation and Reemployment Policies." Speech to OECD symposium on steel in the 1980s, February 27–28. Reprinted in European Parliament Document 1-215/80.

Statistical Office of the European Communities (EUROSTAT). 1974–1986. *Iron and Steel Yearbook*. Luxembourg.

——— 1983. "Aspects Sociaux de l'Emploi dans l'Industrie Sidérurgique (CECA): 1974–juin 1983." *Emploi et Chomage*. September 30.

——— 1984. "Aspects Sociaux de l'Emploi dans l'Industrie Sidérurgique (CECA)." *Emploi et Chomage*. May 11.

Stegemann, Klaus. 1977. *Price Competition and Output Adjustment in the European Steel Market.* Institut für Weltwietschaft geben vor Herbert Giersch, no. 147.

Tsoukalis, Loukas, and Robert Strauss. 1987. "Community Policies on Steel, 1974–1982: A Case of Collective Management." In Yves Meny and Vincent Wright, eds., *The Politics of Steel: Western Europe and the Steel Industry in the Crisis Years, 1974–1984*. New York: Walter de Gruyter.

Tversky, Amos, and Daniel Kahneman. 1986. "Rational Choice and the Framing of Decisions." *Journal of Business* 59 (October): S251–S278.

Vranken, Martin. 1986. "Deregulating the Employment Relationship: Current Trends in Europe." *Comparative Labor Law* 7 (Winter): 143–165.

Yemin, Edward, ed. 1983. *Workforce Reductions in Undertakings*. Geneva: International Labour Organization.

Index